MW00353389

Never Done

Never Done

A HISTORY OF WOMEN'S WORK
IN MEDIA PRODUCTION

Erin Hill

RUTGERS UNIVERSITY PRESS
New Brunswick, New Jersey, and London

Library of Congress Cataloging-in-Publication Data
Names: Hill, Erin, 1977– author.
Title: Never done : a history of women's work in media production / Erin Hill.
Description: New Brunswick, New Jersey : Rutgers University Press, 2016. | Based on
the author's dissertation (doctoral)—University of California, Los Angeles. | Includes
bibliographical references and index.
Identifiers: LCCN 2016003236 | ISBN 9780813574875 (hardcover : alk. paper) | ISBN
9780813574868 (pbk. : alk. paper) | ISBN 9780813574882 (e-book (epub) | ISBN
9780813574899 (e-book (web pdf)
Subjects: LCSH: Women in the motion picture industry—United States—History—20th
century. | Motion picture industry—United States—Employees. | Sex discrimination in
employment—United States. | Sex role in the work environment—United States.
Classification: LCC PN1995.9.W6 H545 2016 | DDC 384/.80820973—dc23
LC record available at http://lccn.loc.gov/2016003236

A British Cataloging-in-Publication record for this book is available from the
British Library.

Copyright © 2016 by Erin Hill
All rights reserved
No part of this book may be reproduced or utilized in any form or by any means,
electronic or mechanical, or by any information storage and retrieval system, without
written permission from the publisher. Please contact Rutgers University Press,
106 Somerset Street, New Brunswick, NJ 08901. The only exception to this
prohibition is "fair use" as defined by U.S. copyright law.

Visit our website: http://rutgerspress.rutgers.edu

Manufactured in the United States of America

For the studio girls

CONTENTS

ACKNOWLEDGMENTS

I thank all of the media professionals I've learned from over the past fifteen years in my hybrid role as media professional-academic. I'm particularly grateful to those friends and former colleagues who have sat for interviews, answered stray questions about union affiliations and interpersonal politics on the set, offered stories from their own careers, connected me with additional interview subjects, and in various other ways helped me understand different aspects of production, past and present.

Thanks to John Caldwell for his mentorship and encouragement, as well as his own foundational research, which helped give form to many of my ideas about production cultures. From John, I learned it is possible to produce good work while remaining a compassionate, ethical, feminist scholar and teacher. Thanks as well to the other UCLA faculty who encouraged me to look at subjects of study from as many angles as possible and gave me to the tools necessary to ground interdisciplinary work. Special thanks to Kathleen McHugh, Stephen Mamber, and L. S. Kim for their invaluable assistance and support as members of my doctoral dissertation committee.

I'm also very grateful to my editor, Leslie Mitchner, for her guidance and assistance over the past few years. I was still learning how to "pitch" myself to editors when we met and connected over this work. I feel very fortunate to have worked with an editor who "got it" right away. Thanks also to Annalisa Zox-Weaver for her editorial assistance.

Thanks to the librarians and staff of the USC Warner Archive, the Margaret Herrick Library, and UCLA Special Collections. Special thanks to Haden Guest and Sandra Joy Lee, Ned Comstock, Barbara Hall, and Jenny Romero, who pointed me in the direction of much of the best archival material discussed within.

I'm grateful to the many friends who have lent their support throughout this work. In particular, Jody Rosenzweig, Jessie Matekunas, Veronica Becker, David Lyons, Karyn Bosnack, Shannon Bates, Brett Fenzel, and Mike Stern. And a special thanks to those friends who are fellow academics and listened and read the most, particularly Maya Smuckler, Jonathan Cohn, Mary Samuelson, Jennifer Porst, and Lindsay Hogan. And to Miranda

Banks, for offering encouragement and advice from her experiences publishing her own excellent book, *The Writers,* with Rutgers University Press. Thanks to other scholars working in this area who shared their insights through conference panels, workshops, and publications.

Finally, thanks to my family. I'm especially grateful to my parents, Tony and Anne Hill, for their love and support over many years spent researching and writing this book. Thanks also to my aunt, Martha Hoppin, who served as one of my earliest readers, cheered me on, and was a resource on the subject of academic publishing. And to my uncle Terry Hill, who acts as unofficial publicist among my family members. Thanks to all the other Hills, Hoppins, Freemans, Johnsons, and Lees. Thanks to my grandfather, John Hoppin Jr., who demonstrates daily the importance of being inquisitive, progressive, and engaged with the world at any age. And finally, thanks to my grandmothers, Mary Ellen Hill and Lillian Hoppin, and all the women in my family, from whom I learned to value my own work and the work of other women.

Never Done

Hundreds and hundreds of experienced girl operators are employed in the factory

Fig. 1. Women film workers at the Selig Polyscope Company's Chicago factory. (*Motography*, October 1915.)

Introduction

> Though women had produced, directed, written, and acted in movies since 1910, by 1973 women did not make movies. They *acted* in them. Major female film stars did not have their own production companies. Women were not heads of networks. They did not run studios. From the 1920s to the 1970s, only a handful produced films. Almost no women directed, they did not run cameras or sound, they did not carry equipment or shout into walkie-talkies. No one said women couldn't do these things. It was assumed. It was the way things were.
>
> —Mollie Gregory, writer/producer, 2002

There's a myth in Hollywood—reflected in both its cultural works and its own internal production culture—that women did not participate in much of film history except as actors or, more rarely, as screenwriters, because they were pushed out from behind the camera in the early years of film-making, managing to return decades later only amid nationwide equal rights activism. This oral history was recounted to me when I entered the film industry as a producer's assistant in 1999, and was repeated by the casting directors I began interviewing about their work in 2004. The same narrative framed many of the industry-authored "Women in Hollywood" books or special issues of *Variety* or *Vanity Fair* published in the 2000s and early 2010s, referencing this bleak past only to dismiss it with a statement about women's progress along the lines of "Today women are working in every area of media production." And, until relatively recently, the focus of

1

most scholarly work on the subject of women in production—on female writers, stars, and silent-era directors—rehearsed a similar story in what it left out.

The Women Who Weren't There

Like many myths, this one is based in fact. With very few exceptions, women did not direct or produce films for major studios from the 1920s, when previous advances in elite creative fields were halted, until the 1970s, when a handful fought their way back to directors' chairs, often under the banner of the women's movement. And, to be sure, the powerful women who survived in the studio system—writers like Frances Marion and stars like Bette Davis—were powerful in spite of the structural inequities that undermined them throughout their careers. These female movie makers—the early, pushed-out pioneers and the rare, studio-era survivors—are important, and their achievements are worthy of all due recognition and attention. But they weren't the only women working in Hollywood in the twentieth century. Their achievements, though remarkable, aren't the sum total of women's contributions to the first century of filmmaking in the United States. And, for better and worse, theirs isn't the only professional line from which the generation of women currently at work in Hollywood descends.

Fig. 2. Women working in the MGM editing room, circa 1921. (Courtesy of the Academy of Motion Picture Arts and Sciences.)

When the earliest American films were produced in the mid-1890s, it was women—the wives of Thomas Edison's male employees—who hand-colored them, frame by frame. Only a few years later, in 1897, a *Scientific American* feature depicted dozens of women bent over spools of film in the American Mutoscope Company's drying and retouching room under the watchful eye of a male supervisor. Like other growing film companies in the 1910s, the Selig Polyscope Company employed "hundreds and hundreds" of "deft-fingered girls" in its Chicago laboratory to cut apart negatives and to patch, retouch, tint, and splice film. This practice was common at the film companies that were, by the 1920s, emerging as major studios in Los Angeles—such as Famous Players–Lasky (which would become Paramount Pictures)—where each film print was projected onto a tiny screen by a "girl operator," and women did all the hand-splicing in the "sample copy" room where films were assembled.[1]

Other women's sectors that sprang up at film studios followed a similar pattern of growth—from a few women working in ad hoc arrangements (which afforded them some professional mobility to nearby male-dominated jobs) to dozens or hundreds of workers in dedicated departments populated almost exclusively by women and often managed by men. A 1920s

Fig. 3. MGM negative cutters, 1950. (Courtesy of the Academy of Motion Picture Arts and Sciences.)

photograph of the Paramount costume department perfectly depicts the system of feminized labor that had, by then, developed to undergird studios' factory-style production methods. Dozens of women in dark-colored dresses are hard at work, their eyes cast down at their sewing; at the center of the frame, their boss, production designer Howard Greer, looks directly at the camera. By the 1930s, an entire floor of the MGM costuming department was reportedly devoted to hand embroidery and beading alone and was populated largely by immigrant women, much like factories in the garment industry at the time. In the 1930s, it was women, working in a similar light manufacturing capacity, who made it possible for Walt Disney to produce his first feature-length animated film on a reasonable schedule and budget. He hired female workers for all of *Snow White*'s inking and painting—the stage of cel animation in which animators' pencil sketches are drawn onto sheets of transparent celluloid with ink and then filled in with paint. Thereafter, the role was so associated with women that the new ink and paint department at Disney's Buena Vista studio was nicknamed "the nunnery."[2]

Women worked in service professions all over studio lots in the 1930s and 1940s. They served meals at studio commissaries, were maids to studio personnel, taught and cared for child actors in studio schools, and provided medical care in studio hospitals and infirmaries. The women who produced and maintained the sea of paperwork on which each production floated formed the largest population of female workers and were arguably most important to the studios' daily workflow. A predominantly female clerical workforce typed and distributed every treatment, outline, and draft of every film in production, as well as most of the notes, memos, and purchase orders that circulated around them. Women administered studio offices as secretaries and looked after the personal lives and emotional needs of executives and major creative personnel. They also filled many of the lower ranks of departments that guided production by means of paper planning; researching productions in studio reference libraries, processing incoming books and plays and evaluating their filmic potential in story departments, typing actor lists in casting departments; and mailing pictures and answering fan mail in publicity departments. Through their collective efforts, managing the flow of paper scripts, records, and other communications—and the details and noncreative work that swirled around creative endeavors—these women were the fuel of Hollywood's large-scale, industrial production process.

In the 1950s and 1960s, studios downsized planning departments by transforming executive-level positions to freelance, middle management, or crew jobs with reduced cachet and increased service components. Predictably, men abandoned these fields, and women streamed in to fill the void. And so, by the 1970s and 1980s, while women like Elaine May, Amy Heckerling, Paula Weinstein, and Sherry Lansing were (re)paving the way to elite creative fields, women who had begun as negative cutters, secretaries, "script girls," and casting assistants were fighting their way into mid-level jobs as junior story executives, production coordinators, editors, casting directors, and publicists. However, because women in the professional sphere were still understood primarily in terms of their gender and only secondarily in terms of their individual talents, skills, and competencies, the low pay and gender stigma associated with their former sectors followed them into these new fields. So, just as script clerks in the studio era were redubbed "script girls," in post-studio Hollywood, women who worked in development (the latter-day equivalent of the story department) were patronizingly nicknamed "d-girls."

On entering these fields, women faced the same implicit gender-related expectations they had faced in their old jobs, where it was presumed certain tasks came naturally to them as women: that on top of their official responsibilities, they add value by "giving good phone," ordering the sandwiches at lunch time, charming and socializing with clients, cheering male colleagues and superiors on, and framing their own workplace contributions with acts of conspicuous, performed femininity so as not to be perceived as bossy, mannish, or bitchy—the worst possible sins for a woman. If women succeeded, it was as part of the team, and credit for their individual contributions was often assigned elsewhere. This was the price of entry into what had been male-dominated positions, paid in exchange for tolerance of women's presence in a workplace that men understood (and often still do) as theirs.

Women were never absent from film history; they often simply weren't documented as part of it because they did "women's work," which was—by definition—insignificant, tedious, low status, and noncreative. In the golden age of Hollywood, women could be found in nearly every department of every studio, minding the details that might otherwise get in the way of more important, prestigious, or creative work (a.k.a. men's work). If film historians consider the classical Hollywood era's mode of production a system, we ought to consider women this system's mainstay, because studios were built on their low-cost backs and scaled through their brush and keystrokes.

In twenty-first-century film and television production, jobs in script supervision, casting, and publicity are still held by a predominantly female workforce and tacitly understood as women's work. Other more gender-integrated jobs, such as producers' or directors' assistants, retain the stigma of having been women's work in the past, an association that contributes to continued low pay and poor working conditions. The same de facto occupational segregation that links women to certain types of media production work effectively dissociates them from others, thereby perpetuating male domination in fields with the greatest prestige and power, the most creative status, and the highest incomes.

There's no such thing as The Truth, especially not in a place like film history, which exists in the mind of the historian. But some things *are* true for me as a woman, a media scholar, and a practitioner working in the industry I research. This history reveals one such truth not yet universally acknowledged in Hollywood: that women's contributions to film history have been vast, important, and ongoing from inception to present, even when their names didn't appear above the titles—or in the credits at all. Simply put: female workers, so often segregated and devalued under the studio system, should not suffer the same fate in media history by being considered only for what they *could not* do as casualties of unjust gender politics. Examining the types of work women *could* and *did* do in the wake of sex segregation reveals their agency—both in their own careers and in their industry's history—and helps frame an understanding of contemporary gendered labor. The stakes—pay, credit, workplace identity, and so forth—are too high to leave the past in the past. And anyway, for women, the past is always present—a reality reflected in every chapter of this book.

The Women Who Were: Feminized Labor as Women's Work

The feminization of labor is a process that has taken place, sometimes recurrently, in subsectors of most major American industries over the past two centuries. During such periods, women have served as "a reserve army of labor," to be hired whenever shifting industrial circumstances demanded a workforce that would accept lower wages and poorer, more precarious working conditions than the existing one.[3] They were hired for the so-called scutwork—the most routine, tedious, menial, and repetitive jobs in the factory or office, performing work often involving light machinery such as sewing machines and typewriters. Once women were on the job, the work and workplace changed further, replicating the era's wider societal norms and tacitly demanding that women adapt to both the work and the norm-enforcing workplace culture. A full explanation of feminized media labor, or "women's work," as it is termed throughout this book, must therefore include not only the explicit, managerial motives for hiring women, but also these implicit, gender-based expectations and the shape the work took as women complied with them.

As one study explains, whether directed at employers, clients, or in support of a business enterprise, "the notion of a service relationship is central to female-dominated occupations," yet typically, the necessary skills or competencies for that service are not formalized in any way, such that "it is even difficult to name them" beyond reference to "'soft,' or 'social' or 'non-objectifiable' competencies."[4] Frequently, service requirements for these female-dominated professions (for example, in hospitality or sales) include *emotional labor*, a category identified by the sociologist Arlie Hochschild, in which the job compels employees to display organizationally desired emotions while suppressing those not desired, "in order to sustain the outward countenance that produces the proper state of mind in others" (for example, a flight attendant's service with a smile).[5] Emotion work and service were—and still are—more often expected of women, based on a set of age-old assumptions about their "natural" qualities and skills as pleasers and domestic caregivers. As Vicki Mayer explains, emotion work is considered a gendered category of labor because when women do emotion work, "they achieve membership in a group identity," while when men do it "it is likely to be seen as an individual trait."[6]

Though personnel files characterized them primarily as typists, the secretaries who assisted studio moguls, executives, producers, and other high-ranking personnel did not succeed or fail on the basis of their typing skills; rather, their success on the job depended on such non-objectifiable skills as tact, charm, and interpersonal sensitivity, all of which they applied to defuse or manage the emotions of the workers around them, absorbing the anger of high-status employers, soothing the egos of their underlings, welcoming their guests, and, in many cases, fending off sexual advances with a smile. These women also displayed conspicuous acts of gendered performance, thus reinforcing cultural notions of static, polar genders.[7] Workplace and larger cultural norms tacitly demanded that women dress and behave in overtly feminine ways. Even at higher levels, there was a narrow range of acceptable women's behavior. A male director who operated in dictatorial fashion (as was the case with Cecil B. DeMille and Alfred Hitchcock, who consciously used personas to intimidate underlings) might be praised for his commanding presence, while a dictatorial female director, if one had existed, was more likely to be criticized for sacrificing her femininity.

I am not the first scholar to incorporate concepts of emotional labor and performativity into the study of film production, but I am perhaps the first to apply them so broadly to reframe existing film history. Such an effort is not always easy or tidy; I wasn't alive when most of my subjects were at work, and few of them are still around to discuss their experiences; yet I see many parallels in my own early experiences as a media industry intern, receptionist, and later executive assistant to film and television producers, writers, and talent managers. This sense of rapport, or, more precisely, over-rapport with my subjects is valuable in many regards. Such empathy can also be treacherous in its potential to lead one to narrativize the events of the women's lives, accepting their memories as fact without taking into account the inevitable distortions that intervene between reality and the production of memory, or identifying with the women in question over their male colleagues. Yet acknowledging gender performativity, service, and emotional labor as compulsory components of women's work is essential to accounting for female workers' lived experiences and their significance to studios. Thus, mindful of my own biases and the fact that I am engaging in an act of performance myself, I attempt to, as Jane Gaines puts it, "tell these women's stories without telling them."[8]

Scope: A Transhistorical Approach

Scholars both in and outside of media studies have characterized women's work as "invisible," for lack of a better descriptor.[9] Yet it is perhaps more accurate to say that people overlook women's work because of their own gender biases—conscious or unconscious. Most people can recognize inequality when they view it removed from their own personal context— their values and beliefs about privilege or lack thereof—as, say, in an episode of the television show *Mad Men*. After observing sexism writ large in that distant, 1960s setting, viewers may be more inclined to recognize subtler forms of inequality when they go to work the next day.

History has the potential to provide readers a similar kind of critical distance. Laying out the entire, 120-year case of feminized labor in Hollywood reveals how conceptions of women's work have organized and shaped women's participation in media production at all levels, across film history. If we can recognize women's work and notions thereof as essential structural elements of the film industry in the past, we—as scholars, producers, creators, viewers—can better identify their remnants today and better comprehend both women's continued struggles and their significant achievements in contemporary Hollywood. For this reason, the intention of this work is transhistorical in nature—a study of a contemporary issue through historical roots. The aim is not to detail every historical era the book encompasses—many complete histories have been written about each decade discussed here—but rather to explore key moments within these eras in order to do justice both to the continuity and significance of women's work throughout these 120 years, and to reveal how the logic surrounding the concept of women's work continues to operate despite changing times and circumstances.

Today, the term *below-the-line* is used to distinguish workers with fixed salaries, distinctly separate from creatively or managerial important *above-the-line* workers with negotiable salaries such as writers, directors, producers, and stars.[10] However, making movies requires not only craft and technical labor (to build sets and run cameras), but also the work of developing film stories and planning production. Though many of the women in this book might be considered to have worked below the line, this classification overlooks many others, because, for example, secretaries were usually considered parts of studio overhead operations rather than members of any particular production. The same holds true for most of today's development workers, who work on multiple productions at once, though only a few production executives may be credited on the finished film.

Leo Rosten coined a more suitable label—*movie workers*—to describe the "anonymous people who swarm over the sound stages, the lots and the offices wherever pictures are fabricated," and who lead ordinary lives away from the *movie makers* (directors, producers, writers, and stars), who were creatively powerful forces in their industry.[11] As with Rosten's movie workers, most of the women in this book lived and worked in Hollywood, but were not "of Hollywood"—they were not members of its circles of power.[12] Much of the book focuses particularly on women in clerical fields, not only because the clerical branch of feminized labor was one of the largest at studios, but also because clerical workers were present in some capacity in nearly all departments and thus offer a variety of useful perspectives from which to view the system as a whole. Exploring this family of women's professions provides a clearer sense of the industrial logic that underwrote all such sectors.

This work owes a debt to the long history of research into women in Hollywood during the silent and early sound eras.[13] Since I began my own inquiries, more excellent work has been produced on women in film history, and particularly women with some degree of creative power in the silent or studio eras. Scholars such as Cari Beauchamp, Jane Gaines, Shelley Stamp, Kay Armitage, Amelie Hastie, Hilary Hallett, and Lizzie Francke, to name just a few, have examined the lives and work of women in early film—from important individuals like Frances Marion, Lois Weber, and Alice Guy to groups of women directors, screenwriters, actors, as well as aspirants to those fields.[14] Still more recently, the Women Film Pioneers Project has begun consolidating bibliographies of existing sources of women in early film and expanding inquiry into some of the very professions this work considers.[15] Perhaps most significant for this project, recent books by Karen Ward Mahar and Mark Garrett Cooper trace the careers of women directors in the silent era and link the narrowing of their professional prospects to women in other areas of production.[16] However, though these studies include feminized labor sectors in their explanations of women's changing professional fortunes, they place primary emphasis on the masculinization of elite creative, technical, and managerial fields (for example, directing, cinematography, producing), and the resultant contraction of the ranks of high-status female directors, producers, and so forth. Taking a different tack, this book attends to the other side of the equation to show that as studios excluded women from certain work sectors, they inevitably allied them with others on the basis of gender. The scholarship of Cooper, Mahar, Beauchamp, Francke, and others serves as a foundation for this history of women in feminized sectors. Many chapters engage directly with these feminist film historians by reexamining their valuable scholarship to further demonstrate how the workplace fates and identities of these two groups were linked. Still, the focus here is on the female movie worker rather than the female movie maker.

Materials and Methodology

Media production is neither a rigid monolith nor a completely idiosyn-cratic, individualistic practice, but rather a *soft system* bound by interre-lated action, whose component parts are often individual human beings with continually shifting frames of reference, continually co-constructing as they go.[17] Women's work within such soft systems was a collaboration between the workers themselves and the forces—industrial, economic, and sociocultural—that acted upon them. As such, women's work must be viewed from the perspective of both the system—the structures that produced gendered understandings of labor—and the individual—the experiences of workers themselves and how they negotiated, resisted, and otherwise co-created their professional identities.[18] Such a top-down-meets-bottom-up methodology requires analyses of a variety of historical and archival materials, industrial, cultural, and economic. Many of the methods John Caldwell and others have used to examine contemporary produc-tion are translated here, to aid to an analysis of historical subject matter.[19] Triangulating among the various forms of disclosure offered by accounts of these workers—from their own statements to the way they were recorded in trade and corporate documents—begins to reveal the "shop floor practices" so often hidden within official or formal documents, or stories told after the fact for posterity or as professional self-representation.[20]

Because women movie workers' contributions to production were gen-erally deemed less important than those of their male and movie maker colleagues, studio archives seldom held on to materials relating primar-ily to these workers.[21] Documents that do remain were usually preserved because they could help tell the story of an important actor, writer, director, producer, or film text. In "Reevaluating Footnotes," Radha Vatsal argues for the excavation of footnotes as sidelined texts, for both their information and the "contradictions and equivocations" that are often discarded into them, pointing out the footnote's potential for self-reflexivity and raising new questions about the writing of history.[22] This work frequently seeks out subjects in peripheral ephemera documents and in the footnotes and margins of other people's histories.[23] The location (or lack thereof) of evidence serves as evidence in itself, and helps to tell the story of women workers through the ways they were variously remembered, classified, and erased in archives. Like many "lost" or "missing" films of the silent era, much of this history may truly be gone, but there is also much that may be rediscovered.

The map—literal and figurative—is an important part of this evidence. An ideologically loaded practice, mapping has commonly documented historical change through the eyes of that change's chief architects: political leaders, military generals, and other major historical figures or "great men." Maps (of geographic space, production hierarchy, and workflow) also played an important role in the scientific reorganization of labor practices after the industrial revolution, and they were much beloved by early studio managers, who used them to represent their own agendas and to declare their ownership. Such mapping schemas were means of women's exclusion at studios, which were styled after modern cities, complete with cities' gendered spaces. However, as Denis Wood explains, maps record not "what was," but rather, the interests of the people who drew them, showing "*this*, but not *that* . . . this *way* . . . but not *the other*," and thus embodying their authors' biases and values.[24] Maps and related discourse around the building of studio lots reveal studios' interests through what they showed and left out. In indicating where and how female workers were (or, more often, were not) represented, such maps suggest where they were being (re)located in these developing organizations.

Anecdotes of, by, and about female workers in their letters, studio newsletter columns, memoirs, and biographical accounts serve as correctives to the public, top-down modes of self-representation schematized in the studio maps other documents. Anecdotes connect the individual to the system, representing specific, local, and personal accounts that both complicate and contextualize those depictions of the various production systems as naturally occurring or "the way things are." As such, the anecdote serves as both countermap and companion to the map, providing an unofficial vocational tour through the studio spaces maps purport to describe.

Studio house organs, as well as memoirs, oral histories, and other after-the-fact writings, have been the main sources for such anecdotes. Such writings are subject to distortions. House organs—monthly newsletters written for and by employees—present a picture of studio life that is necessarily aligned with and shaped by management's goals, and must be deciphered as such. Workers who wrote newsletter columns often did so in the interest not of accuracy or fact-based reporting, but of managing their disclosures about themselves, their departments, and their colleagues to serve their own occupational aims or to reinforce perceived socioprofessional boundaries and rules. The anecdotal bits and pieces newsletters collected are most useful when understood as evidence of their work culture's values and norms—how workers framed their work for each other, under the eyes of their employers—rather than as ironclad evidence of particular practices. Because much of the work of this book concerns interpersonal interaction and socialized gender roles in the professional sphere, the gossip and inside jokes related through newsletter columns provide necessary context for the areas of women's work under discussion. However, these newsletters occasionally include more descriptive reportage in larger features devoted to outlining studio organization or the structure of departments for the newsletters' movie worker readership. Cross-referencing these accounts with those in trade papers, studio records, and other historical accounts helps elucidate studio structure and practices.

Memoirs, oral histories, and other accounts written long after the events under consideration took place are subject to misrecollections and to authorial self-aggrandizement, especially in the case of former movie workers such as secretaries and script supervisors whose work was so often under-rewarded and overlooked. As Amelie Hastie explains in *Cupboards of Curiosity*, while women have been written into and out of media history, they have also written themselves in through memoirs and other works. Different chapters in this book attempt to balance the accounts these writers give of people and events long past (through corroborative primary and secondary resources) as well as to balance existing histories' accounts of the women, typically in the biographies of their employers, through their reflections of how it felt to do such work and how they viewed their roles years later. Drawing on Walter Benjamin's work, Hastie explains how "a shock experience in the present lays groundwork for excavation of memory in future," in the way that "a present moment reaches to the future, and the present that future becomes reaches back to the past."[25] In my interviews with women media workers, as well as my research into women's history, I've been struck by how often women's accounts of and feelings about the work they did and the conditions they accepted in feminized sectors change even a few years later, when the rewards implicitly promised for such work (acceptance, promotion, and so on) fail to materialize. Thus, interpreting these memoirs has meant engaging their inevitable embellishments not just to avoid reproducing historical inaccuracies, but also to better understand the subjective truths they often reveal about people who lived through and beyond "film history" and perceived themselves as being forgotten while those they served were remembered. These accounts are perhaps most valuable in their descriptions of work in feminized sectors: what such work consisted of, and what it felt like to the worker. Interpreting their descriptions of work on film sets often requires analysis not only of what is said but how it is expressed, as in the case of script supervisors, whose constant reference to "worry" and "care" when describing their roles provides insight into the overriding function such work served in production.

It might be easy for these collected anecdotes—gathered from old biographies, house organs, studio histories, and so on—to remain just anecdotes; that is, isolated incidents that testify only to the particular personalities or events they concern. However, discrete anecdotes begin to cohere when organized by the framework of *creative service*—a term that accounts for the various components of women's work by connecting the more straightforwardly feminized duties that characterize so many women's jobs (detail work, routine, mechanized tasks) to the forms of feminine performance and emotional labor women deployed to support the moviemaker or creative collective (service, smiles, pleasing manners, and so on). The concept of creative service operates as a spectrum along which common characteristics of feminized film work may be identified and linked through their most essential shared purpose, revealing the overarching industrial logic that united seemingly disparate feminized sectors. Scattered stories related by or about individual workers thus coalesce, assembling a picture of what women at studios actually did and how their labor benefitted Hollywood's larger, studio-era industrial project.

This methodology functions not to articulate the specific contributions of one worker, but rather to comprehend systems of creative production and how workers participated and served collectively in that process. Thus, this book contributes to authorship studies only in the sense that it complicates understanding of the large studio production process and credits feminized labor sectors with combined contributions to American media history. Its approach can exist alongside and enhance our understanding of major creative figures and their authorship by countering the notion that Hollywood was shaped exclusively by a series of moguls and auteurs with the accounts of the women who worked for and around them. The book's counter-histories are not meant to supplant the stories of these "great men," to cast their female underlings as the true authors, or to rank them above other contributors to production. Rather, they are offered to suggest that our esteem for great movie makers often acts like a spotlight, plunging the contributions of the less conspicuous into darkness and rendering the great men themselves less interesting by blowing out their most well-known features with harsh, flat light. Such excessive regard simplifies the shadowy complexity on all sides of creative collaboration. Accounts like those of the women in this book can provide balance—a kind of historical fill light—bringing nuanced details into view.

Organization

Chapter 1 first examines the feminization of the American office in the nineteenth century, as growing industries attempted to scale growth through incorporation of scientific management principles, and then follows women's entry into the American motion picture industry in its early years. In contrast to their participation at elite levels in the relatively heterosocial workspaces of film production in the 1900s and early 1910s, women's workplace identity shifted under studios' efficient reorganization in the mid- to late 1910s and early 1920s, when certain aspects of the production process were separated, classified, and relocated (as such) according to studios' logic. After this shift, women were increasingly associated not with creative or craft specialties, but with the clerical sector through which studio management coordinated their efforts.

Chapter 2 recounts how, as studios increasingly associated women with established categories of feminized labor from other American industries (for example, domestic service, clerical work), they effectively dissociated them from high-status creative fields in which they had found much early success (for example, directing, producing). These associations took on a spatial dimension as female workers began to occupy very specific places on studio lots—and thus in the minds of studio workers—while their employment in others dwindled. Self-promotional "studio tours"—short films produced by studios that purported to give backstage views of their lots—guide the chapter's alternate tour of women's work at studios as it was construed by management, which expands the book's scope to include immigrant workers and people of color, male and female, who occupied the lower tiers of studios' labor hierarchy alongside the white women who populated feminized sectors.

Chapter 3 focuses on the clerical branch of women's labor, outlining how clerical workers were distributed across the lot, connecting the three main functions of studio organizations: overhead operations, planning, and production. Moving beyond the basic duties assigned to clerical workers in studio personnel documents, the chapter also examines how sociocultural expectations of women in the classical Hollywood era extended their work outside of those parameters by implicitly requiring them to perform femininity and sexuality on the job.

To account for the implicit politics involved in playing these feminine roles, chapter 4 further focuses on one subgroup of clerical workers: secretaries and assistants to movie makers. Where earlier chapters rely on official studio-authored accounts of women's labor, chapter 4 reveals women's roles primarily through unofficial, private accounts of the workers themselves, constructing case studies of five notable secretaries from the studio era. This chapter introduces the concept of creative service, using it to link and explain different secretaries' work through common traits. These accounts illustrate the high level of creative and managerial agency some women achieved despite their industry's structural inequalities, revealing they had greater significance than their marginal representation in existing film histories implies.

In a larger sense, all women's work at studios can be understood as a form of creative service because all women's sectors fulfilled the same larger purpose of absorbing routine tasks and unwanted emotion around men's creative process. This dynamic can be seen most clearly in formerly masculinized or gender-neutral production specialties that later shifted to become women's work. To explicate this feminization process and its downstream effects on media history, chapter 5 traces the origins of several contemporary women's professions with roots in women's clerical labor. Starting with the production specialties that feminized earliest and are most closely linked to feminized labor—development labor, script supervision, and publicity—the chapter then examines two outliers whose origins and gendered aspects are less apparent: editing and casting. Considering editing and casting alongside unmistakably feminized jobs reveals how the logic that guided feminization in the studio era continued to operate decades later and spilled over into nearby roles.

Once unraveled, the story of women's media professions and their feminization highlights the global function that gendered labor served in the Hollywood studio system—and continued to serve long after this system ceased to exist. Feminization has greatly affected the media industry's subsequent evolution in terms not only of gender integration, but also the way workers identify within professional groups, their creative practices, and the products that result from them. The book's epilogue provides a snapshot of the studio secretary's present-day counterpart—the producer's or director's assistant—to illustrate how notions of women's work written into production culture and job descriptions during the studio era continue to be circulated as industrial mythology today—the only real order or continuity in otherwise chaotic, post-studio production cultures.

1

Paper Trail

Efficiency, Clerical Labor, and Women in the Early Film Industry

Women have only been allowed to encroach on areas that have limited power. We can type up our own lists and make a deal at the same time.

—Jane Jenkins, casting director, 1991

The cultural logic that determined which film production jobs would be assigned to women developed before the advent of moving pictures, during nineteenth-century experiments with the scientific management of production and debates over whether and how to employ the groups of middle-class women that had emerged as a potential new workforce in urban centers. The answer, it turned out, was to sit them down at typewriters as part of a larger strategy to subsidize and control the massive growth of many American businesses after the Civil War. Once it took hold, the notion of women as natural clerical workers became nearly impossible to eradicate. The pattern of clerical feminization that arose across American industries in the nineteenth century would go on to shape and sanction film industry production practices in the twentieth.

Women's relatively high level of participation in the heterosocial workspaces of early film production began to shift in the late 1910s and early 1920s, with the "efficient" reorganization of studios and attendant sex segregation and feminization practices under which work was separated, classified, and relocated (as such) to accommodate the studios' aims. According to this logic, the largest feminized labor sector—that of clerical work—emerged as a key component of efficient mass production, facilitating expanded management and cost accounting, reducing labor costs, and absorbing mass film production's lowest-status, most repetitive, least desirable forms of labor on the basis that they were women's work. As it had elsewhere, efficiency shifted the film industry from less formal, more holistic early work systems in which women moved fluidly between different work sectors (presenting a kind of unintended or latent feminism), to a highly structured, rigid organizational model in which management was geographically and conceptually separated from production. Under this logic, women were increasingly identified with neither creative/managerial specialties nor production jobs, but rather with the clerical sector that connected the two (manifesting an overt feminization more in line with the cultural and professional norms of other industries).

Building on Janet Staiger's seminal work on studio organization, this chapter sketches the geography of the early film industry both before and after the onset of full-blown efficiency by following the paper trail—quite literally, the increase in paper processes at studios as they incorporated scientific management principles—that eventually rationalized the use of large, feminized labor sectors.[1] Once introduced, sex segregation's gendered geography became embedded in studio culture, both geographically—written into maps and displayed in promotional materials—and ideologically—in the production workflow and hierarchy of studio labor.

Clerical Labor as Women's Work in the Gilded Age (1850s–1900s): The Pre-mechanized Office and Early Efficiency Practices

Clerical work was so associated with women by the first decades of the twentieth century that female secretarial archetypes such as the long-suffering Girl Friday were already well-established staples of films and novels.[2] Yet in the work of Charles Dickens, George Bernard Shaw, Herman Melville, and other nineteenth-century novelists, the male clerk is a ubiquitous figure while female clerks are almost nonexistent, reflecting the masculinized state of the office through the middle of that century.[3] For centuries, male clerks and secretaries had served as record keepers and letter writers for governments and nobility in Europe and the United States. By the 1800s, small businesses such as banks and insurance companies regularly employed male clerks, bookkeepers, messengers, and copyists or scriveners to create and maintain office paperwork. In these pre-mechanized offices, the boundary between clerical and managerial workers was blurry, and, as Margery Davies explains, a single clerk might master "the entire scope of an office's operations," learning from an employer as an apprentice studies under a master craftsman.[4] An employer dictated his clerk's working conditions as well as his mobility through the business world and, by extension, the class system.

In the wake of the Civil War, businesses expanded from small, competitive firms to larger, more vertically and horizontally integrated corporations with monopolies on major industries such as steel, oil, and meatpacking. To coordinate among geographically dispersed branches of their businesses, growing companies employed new forms of paperwork to track costs, dictate workflow, formalize previously idiosyncratic practices, and enforce more rigid departmentalization and hierarchy. Order was imposed on the chaos of large-scale production through the adoption of the principles of "scientific management of production," first developed by mechanical engineer Frederick Winslow Taylor. Scientific managers studied different workers in order to synthesize their "best practices"—those techniques that evolved over time as most effective—and standardized the steps of their work processes, reorganizing the workplace to minimize waste and maximize profitability. "Efficiency," the term commonly used to signify these collective practices, swept like a fever through countless areas of American life, from steel mills to government offices to private homes. As Sharon Hartman Strom observes, efficiency practices offered a solution to the problems of industrialization, bridging the gaps "between progressive reformers and railroad tycoons, efficiency experts and production managers, consumers and corporations," who put their faith in a common language of balance sheets and systematization.[5]

Scientific managers used new systems of paperwork and computation, appropriating knowledge and decisional authority from previously self-governed workers like machinists and handing it over to managers in planning departments who were physically separate from production.[6] Historian Lisa Fine explains, "Clerical workers do not produce the good demanded" in a manufacturing context, "but they produce an intermediate good that economists call clerical output."[7] Efficiency elevated clerical labor in importance as clerical output became a bridge between the "brain" of management planning and the "hands" of production, linking different departments and employees through marching orders without which they could no longer function.[8]

All of this paperwork (and the additional bodies needed to process it) posed its own threat to cost effectiveness, so scientific managers also studied clerical workers' best practices and then rationalized, standardized, and separated their work processes, redistributing their decisional responsibilities to managers and delegating many other responsibilities to subordinates, each specializing in a narrower share of the work process. New clerical designations included timekeepers, payroll clerks, accountants, billing clerks, ledger clerks, cost clerks, key punch machinists, requisition clerks, shipping and receiving clerks, stenographers, typists, Dictaphone transcribers, and switchboard operators.[9]

Efficient work systems also employed an array of newly invented organizational and labor-saving technologies, from filing cabinets and index cards to stencils, mimeographs, and Dictaphones. The typewriter came into use in the 1870s and, together with stenographs, mechanized the writing process and eliminated the need for a copyist's handwriting skills. The profession of steno-typist—the entry point for most women in turn-of-the-twentieth-century offices—was not unskilled; it required training in shorthand and typing. Nonetheless, stenography and typing technologies broke down the process of copying documents into multiple, mechanized steps that might be completed more quickly and with less variation between operators. The operator became a conduit through which information passed—unaltered—from sender to receiver, rather than the active, human participant the copyist had been in message creation.

Efficiency's benefits to management were manifold: unskilled or semi-skilled laborers demanded lower wages than skilled workers, had less expectation of promotion beyond the clerical sphere, and were easy to train and thus easy to replace if they quit or were fired. As a result of this change in circumstances, extant male workers left clerical fields in large numbers from the 1880s to 1910s, while young men embarking on careers in business looked elsewhere for the pay, training, and advancement potential they had once enjoyed as clerks. Men who did remain typically did so in positions of higher skill and status, such as manager, accountant, or supervising clerk. Meanwhile, a shift from agrarian to urban life brought young, single women—previously engaged in their families' small businesses, farms, or homes—to cities.

The Feminization of the Office

According to nineteenth-century Victorian ideals, women's "natural" sphere was the home, and the only acceptable types of employment for a woman were roles related to the domestic sphere, say, as servants or teachers. However, amid industrialization and urbanization, more working-class women needed work than could be employed in domestic service jobs, and the ranks of literate, educated, middle-class women outgrew available positions in culturally acceptable women's fields such as teaching and nursing.[10]

Debates about women's fitness for office work were rooted in culturally held assumptions about the essential characteristics of each gender. Many feared that women's mass entrance into the workplace would threaten the balance of perceived innate skills and qualities that clearly delineated the genders, whereby women occupied the private sphere as the moral, religious, emotional, aesthetic, intuitive complements to rational, logical, ambitious, strong, practical men in the public sphere.[11] Counterarguments held that women would improve workplace morals through many of the same essential qualities of their gender. From these debates later emerged the progressive, feminist ideal of the New Woman, a figure who both sought and symbolized new social freedoms for women. Hilary Hallett explains that the concept of the New Woman came to represent changes associated with women's rapidly increasing participation in work outside the home between 1890 and 1920 as, "for the first time the majority of women, usually those who were white and native-born, experienced work as an endeavor that sent them outside the confines of factories or other women's homes" to offices and department stores.[12]

Employers in need of cheap, disposable labor submitted arguments in favor of women's particular suitability for clerical work. Said one, women had the quickness of eye and "delicacy of touch which are essential qualifications of a good operator," took more kindly to sedentary jobs than men, and were "more patient during long confinement to one place."[13] The association of immigrant and working-class women with "light" manufacturing work (textiles and such) was deployed as evidence that women might work in a "precisely similar" capacity in the rationalized office.[14] Soon after its invention, the typewriter began to be marketed specifically to women. Salesmen used female operators to demonstrate the new technology on the basis of women's skill at playing pianos and operating sewing machines. Women were held to be nimbler and neater than their male counterparts, whose "broad tipped fingers," an 1880 catalog explained, "do not fit him for a graceful operator."[15] Catalogs and advertisements dubbing female operators "type-writers," further conflated women's bodies with the machines.[16]

Women were paid less—as little as half as much—for the same work, a practice rationalized through the cultural assumption that a woman would leave the office once married for her preferred sphere (the home) and that a man should receive a "family wage"—enough to support his family on his salary alone. Many firms also instituted a "marriage bar," which forced women to retire once they wed, and female workers were typically ineligible for promotion.[17] All of these practices eliminated the threat women posed to male coworkers as well as employers, ensuring that the information they managed would pass harmlessly out of the office when they left for married life. Jennifer Fleissner explains that male clerks learned to avoid close association with the skill of stenography so as not to be permanently marked, as female clerical workers were, "as recipients, rather than originators of dictated information."[18]

With few other options for employment, women clerical workers' numbers grew rapidly at the turn of the century, ushering in a new age in which technologies for mass communication were gendered female and clerical work was feminized. By 1908, the US Census reported nearly half a million women in largely clerical roles related to trade and transportation.[19] The percentage of female bookkeepers, cashiers, and accountants in the United States rose from 2 percent in 1879 to 51.9 percent in 1930, while the percentage of female stenographers and typists rose from 4.5 percent to 95.4 percent of the total workforce during the same period.[20]

Changing Expectations in the Feminized Office

As Kathleen McHugh explains in *American Domesticity*, by the early twentieth century efficiency and scientific management had found their way into the domestic sphere, where women were expected to be visible symbols of leisure, concomitantly rendering their labor less visible.[21] Home economics manuals taught women to be both their household's manager and its labor force through new machines and efficiency principles. At the same time, these manuals instructed readers to disguise the considerable work involved in maintaining the household through—as one domestic engineer put it—"an inordinately delicate feminine appearance and manner."[22] Office labor underwent this makeover in reverse as it became linked with the emotional and maternal values espoused by the nineteenth-century cult of domesticity. Thus, when women took over clerical roles, the expectations of those roles almost immediately expanded to include aspects of feminine performance and emotional labor that, Alice Kessler-Harris explains, "sustained women's roles by extending their home functions to the job."[23]

Business journals and women's magazines alike reconceived the new, feminine office as a home, and the new, female clerical worker as its mother, mistress, wife, caretaker, and hostess. Trade journals reported that women brightened the workplace through their femininity, raised the moral tone (inhibiting swearing and smoking through their presence), and created an atmosphere more like the parlor in the domestic sphere.[24] Articles on women in the business world posited the office as a perfect training ground for marriage, since a woman who distinguished herself as a stenographer or secretary increased her value to a future husband, whom she could serve as "a business companion as well as a wife and a social companion."[25]

As in the home, women in the office were expected to render their labor invisible by virtue of a feminine appearance and behavior. Secretarial guides advocated a style of service in which workers did things without seeming to do them, and, on top of everything else, maintained a "pleasant and cheerful environment."[26] If women's labor posed a threat to Victorian values, this alliance between clerical labor and traditionally feminine traits, qualities, and behaviors was one way to neutralize it.

Women as Secretaries

Though she might disguise her considerable responsibilities under a veneer of femininity in her office-as-parlor, the secretary was a manager in her own right, running her employer's office like the clerks in the early 1800s businesses. Vocational materials from the first half of the twentieth century reflect a hierarchy of desirability, with typist (who merely typed up documents and notes) held in low regard, stenographer (who took dictation in face-to-face sessions with executives) only slightly higher, and secretaries (who provided support to executives) held in highest esteem because of the requisite intelligence and interpersonal competency. As one secretary summed up, "A stenographer . . . is paid to do; a secretary is paid to think."[27]

Self-direction and other more solid managerial abilities were still required for the work because secretaries often oversaw other clerical workers in their employers' offices. Yet the field did feminize, in part, because the potential for promotion from secretarial positions was removed or restricted, and in part, because women demonstrated themselves as ideal candidates for the positions, effectively "sharing" their jobs with their employers. As Davies explains, "In a patriarchal society it was natural that a male employer should give orders to and receive services from his female private secretary," and that, as work was divided between them, "the man should do all the creative, 'important' parts and the woman all the routine, 'unimportant' ones."[28] Where tension might arise from male secretaries wanting credit, promotion, or increased respect in exchange for a job well done, a woman's subordinate position as a man's little helper was posited, gender-normatively, as its own reward.

In fact, women were deemed such natural fits for secretarial work that they were almost immediately expected to expand their roles to include more feminine service and gender performance. As secretaries, it was believed women could put to use the "social gifts" that they used in the home.[29] Female applicants were encouraged to cultivate tact, an even disposition, and "an obedient and slavish type of mind, rather than a vigorous and constructive one," because a successful secretary, it was said, "thinks *with* her employer, thinks *for* her employer, thinks *of* her employer."[30] Instead of supporting an actual mate, the secretary supported her workplace mate in his endeavors, benefitted from his successes and took part in his failures. In the words of one handbook, "Every man needs a woman's tenderness and her pride and faith in his ability, to buck him up in the fight he must make in these days of terrific competition."[31] When hiring a secretary, employers were advised to "select a woman who you think could be married at any time if she chose, but just for some reason does not."[32] However, though great secretaries might become great wives, the reverse was not true, as a married secretary's "primary loyalty had already been given to one man," and her "attitude toward men who come to the office is not at all the same as that of an unmarried woman."[33] The common conflation of secretary and wife eventually gave rise to the sobriquet "office wife" and "work wife."

A 1935 *Fortune* magazine description of an office illustrates the compulsory gendered roles women were assigned as clerical workers:

> The male is the name on the door, the hat on the coat rack, and the smoke in the corner. But the male is not the office. The office is the competent women at the other end of his buzzer, the young ladies chanting his name monotonously into the mouthpieces of a kind of gutta-percha halter, the four girls in the glass coop pecking out his initials with pink fingernails . . . and the elegant miss in the reception room recognizing his friends and disposing of his antipathies with the pleased voice and impersonal eye of a presidential consort.[34]

In describing the office's conspicuous femininity the author invokes spaces of domestic, feminine servitude—the harem is vaguely suggested, and its animal kingdom counterpart, the chicken coop, is referred to directly. Female office workers' success was achieved through their willingness to play such roles.

Female clerical workers began to appear in novels and other cultural works in the late 1800s and soon permeated popular culture, all but eliminating the male clerk character of the Victorian era. In novels at the turn of the century, "girl" typist or secretary heroines embodied the modern age.[35] For some, these characters were also frightening, embodying fears of technology and gender in the new age. Though positive, optimistic characterizations of "type-writer girls" were common in turn-of-the-century novels, so was the notion of the typewriter—and thus the woman who operated it—as mystifying, even supernatural, and often the means through which to contact the spirit world or enact nefarious plans.

Female typists and secretaries were also commonly depicted in the twentieth century's newest popular entertainment: motion pictures. At the same time that images of women office workers began to appear in early American films, notions of women's work were finding their way into Hollywood film production, threatening women's early success in the burgeoning movie industry's elite creative professions by relegating so many of the women who came after them to feminized labor sectors and suitably feminine roles within studio culture.

Pre-efficient Hollywood as Latent Feminist Work System: Women in Early Film Production (1890–1909)

At the turn of the twentieth century, the movie business was in its nascent stages—a long way from its eventual status as a major American industry. The first films emerged in the 1890s from what Karen Ward Mahar calls "the highly masculinized settings of the inventor's laboratory," and new film and projection technology was gendered male just as still photography had been.[36] Far from being mass produced, the earliest films were short, experimental efforts concerning nonfiction subject matter produced by only a few employees. For example, *Edison Kinetoscope Record of a Sneeze* (1894) consisted of one employee in Edison's Black Maria studio (Fred Ott) filmed by another (inventor W. K. L. Dickson) with no script, actors, costumes, or sets.[37] Although it is impossible to verify that women—or clerical workers, for that matter—were completely absent behind the camera during the production of these early motion pictures, existing evidence attests to their absence as a general rule.

Film technology improved rapidly between 1894 and 1904, and demand for longer and, eventually, narrative or "story" films increased. Productions became larger and more complex, yet production staffs were still quite small by the standards of even a few years later, and ran under the direction of cameramen who controlled their own productions from start to finish.[38] Janet Staiger has likened this "cameraman system of production" to the work system of an artisan or craftsman under whom all stages of conception and execution are unified—basically the work of one man.[39] Under this system, women's participation in film production beyond the level of actor was nonexistent; however, outside the physical production phase, the developing movie business was less exclusive, and many emerging movie jobs closely resembled extant women's jobs in related industries.

Photographic manufacturing was already associated with women, as evidenced by the largely female staff displayed in the 1895 Lumière film *Workers Leaving the Lumière Factory*.[40] And though Edison's laboratories were masculinized spaces to be sure, photographic detail work such as the hand coloring of early Edison films was farmed out to the wives of the male workers.[41] By 1897, the American Mutoscope Company employed a staff of women retouchers and negative cutters, a practice that was widespread by the 1900s.[42] Women also served as clerical workers in areas of the business related to—but not falling directly within—the realm of film production. Still photography had existed long enough to necessitate sales and business offices, and because the same companies manufactured film, female clerical workers were important parts of the film manufacturing business. Most notable among accounts of secretaries at photography companies in the 1890s was that of Alice Guy Blaché, who, before her rise to prominence in the American film industry, worked as a secretary at Gaumont in 1895 when it began developing motion picture cameras. Blaché became a director when she "borrowed" a camera prototype and shot what was "arguably the first fictional film."[43] Other early accounts include those of secretaries working in the offices of film financiers, exhibitioners, or the manufacturers of photographic equipment. As Hilary Hallett explains in *Go West, Young Woman! The Rise of Early Hollywood*, by 1900, commercial entertainment was already "one of the largest, best-paying fields open to women without much formal education." Women worked in film exhibition as theater owners and ticket takers—a practice carried through from theater and vaudeville—and as office workers in film sales, rental, and booking offices, which could not function without paper record keeping.[44]

Between 1904 and 1907, directors (often coming from the theater) began earning employment in staging films' increasingly complex, choreographed action before cameras. This "director system of production," as Staiger has called it, took its cues from the production of stage plays, in which the director managed workers on set and had ultimate authority.[45] Where before little if any hierarchy existed among workers (though there might be between management and craftsmen), now there was a pyramid of workers, with the director situated securely at the top; yet formal planning was still virtually nonexistent. Outlines were constructed from a basic plot broken into parts for filming, or from "the casual scenario," which might be as short as a few lines scribbled on scratch paper—and might change according to the director's discretion.[46] This setup meant that "those who actually made the films were craftsmen," functioning as generalized experts, not unlike the Victorian clerks discussed in the previous section, rather than specialists in only a few tasks.[47] Thus, more workers operated under this system, governed by a loose hierarchy and some basic planning, but, important for female film workers who entered the industry in the first decade of the twentieth century, they were still generalists adhering to a holistic view of the process—not separated, standardized, or centrally managed. This arrangement was, then, a relatively informal, small-group filmmaking process from start to finish.

At this time, growing film production companies employed some clerical workers as secretaries in their business offices, and for bookkeeping and other paperwork-related tasks. Edison's New Jersey factories employed female secretaries, as did the business offices of other growing film companies.[48] Some early directors and stars hired private secretaries for correspondence and other areas of their businesses. For example, Charles Higham reports that Jeanie MacPherson was first hired by Cecil B. DeMille at $25 per week as a longhand stenographer before she became the director's longtime screenwriting collaborator.[49] But such workers did not exist on a large scale and were mostly disconnected from physical production.[50]

During that first decade of the twentieth century, as movies emerged as lucrative popular entertainment (because they attracted both male and female customers), the relatively artisanal systems of production at various film companies meant production roles were not as strictly defined as employment in more established, recently Taylorized industries. And because many motion picture directors, writers, and producers were recruited from theater, they brought with them the tradition of "doubling in brass," a term from minstrelsy meaning to carry out multiple jobs when production required it.[51] Also imported from the theater and vaudeville was a lack of formal boundaries between men's and women's labor, because, among other things, male and female workers (actors, artists, and so on) shared the same workspace. As such, when help was needed during early film production, anyone and everyone was expected to pitch in regardless of station or gender.

Hallett explains that by in 1910 women outpaced men in migrating west.[52] Los Angeles's real estate, tourism, and film industries created service and clerical jobs that attracted female workers, many of them young and single. California's passage of women's suffrage, jury service, and legislation establishing an eight-hour day in 1911 and a minimum wage for women in 1917 meant that though gender and, particularly, racial inequality remained widespread in the city, "white women experienced less social stratification and greater legislative protections than most cities." Developing notions of Hollywood and the early success of women there would soon become a powerful lure for a "New Western Woman in full flight from gender norms," drawing ever more women to the state, and to Los Angeles in particular.[53]

Absent a standardized production practice, female workers were expected to cultivate a holistic understanding of the business alongside their male counterparts. Female workers' accounts of their early experiences in the film industry convey a sense that gender played less of a role (relative to industries other than theater or vaudeville) in determining which duties were suitable for which workers. When later director/producer Dorothy Arzner was first hired as a typist for silent-era screenwriter and director William C. DeMille (elder brother of Cecil B.), she was instructed that she should use this vantage point to learn the whole business.[54] Similarly, when director/producer Lois Weber hired Frances Marion as an actor—something

Marion had little interest in—Weber assured her that at most studios, "everyone did a little of everything," and that as Weber's assistant and protégée, "she would work in every stage of production, including in front of the camera."[55] Said Marion of this experience, "During the weeks that followed I skittered around the studio doing every kind of job I could find except emptying the garbage pails."[56] In her first job under Weber, Marion did "whatever needed doing: writing press releases, moving furniture on the sets, painting backgrounds, and mastering the art of cutting film."[57] She also wrote dialogue for extras, learned set and costume design, rode horses as a stunt double, "read scripts and dared suggest changes, and even hauled furniture around to make the sets more attractive."[58] Writer Beulah Marie Dix, who began her career at Famous Players–Lasky, described filmmaking as "all very informal, in those early days. There were no unions. Anybody on the set did anything he or she was called upon to do. I've walked on as an extra, I've tended lights (I've never shifted scenery), and anybody not doing anything else wrote down the director's notes on the script."[59]

In this informal work system, a few women infiltrated such male-dominated fields as cinematography, location scouting, publicity, and even studio management.[60] Many others who emerged as figures of creative or managerial importance in the early film industry similarly ascended from the lower ranks of film companies to roles as writers, directors, producers, and production company owners. As Cari Beauchamp explains, women such as Cleo Madison, Lois Weber, Margaret Booth, and Anne Bauchens found success because "with few taking moviemaking seriously at the time, the doors were wide open to women."[61] And as Hallett argues throughout *Go West, Young Woman!*, as both movies and trade and fan magazines began to cater to women, increasingly perceived as an important or even a primary audience, their narratives reflected Hollywood as a place in which gender norms might be contested. By underlining the examples of successful women filmmakers like Weber and Marion, such films and magazines supported boosterish discourse and encouraged more women to seek work in the movies via the fantasy in which the idea of working in Hollywood was wedded to women's desire for emancipatory change.

However, in the late 1910s, those doors would begin to swing shut. Ironically, the achievements of filmmakers like Weber and Marion probably hastened the end of this fluid, heterosocial work environment. For, with each successful film they made, early female filmmakers increased the viability of movies as an industry by increasing consumer demand. Movies would soon follow other American industries down the path of scientific management, standardization, and sex segregation. As the number of available female workers in Hollywood increased, their agency in production decreased and was channeled into increasingly feminized forms.

The Efficiency Men: Scientific Management in Developing Systems of Production (1910–1915)

Though profitability in the film industry was growing, production practices for maximizing that profitability had not developed apace. The director system of production provided little in the way of advance planning because directors' practices were idiosyncratic. Directors often hired and fired their own staffs and gave them oral instructions rather than written ones; sets, costumes, and props were sometimes borrowed from other sets or rented with no system to track or ensure their return; and purchasing might be carried out through multiple agents with little cost-control oversight.[62] Directors divided their attention between resource management and artistic goals, and did so practically from scratch for each new film. This system sufficed for a time, but by 1910, the increase in production volume made plain its inadequacy.

To expand and meet growing demand, motion picture companies courted infusions of outside capital from investment firms but met with limited success because of the perceived instability of the fledgling film industry and its production practices. This perception was bolstered by the failure of two companies, the Triangle and World Film Corporations, which had attracted investments early on.[63] Like others before it, the film industry looked to the stabilizing concepts of rationalization, efficiency, and scientific management, first to solve its growing pains through cost reduction, and later to secure investments by cultivating the solidity Wall Street investors required through professionalization.[64]

Motion picture trade and fan publications discussed efficiency with increasing frequency in this period, alongside discussions of expansion and relocation of many companies to Los Angeles. Journalists reported with growing enthusiasm on the adoption of efficiency techniques, praising those who implemented them, calling for further systematization of the production process, and arguing the measures were necessary because of the increase in business volume and film length, because "the making of a big feature play requires differentiation of several crafts."[65]

By the 1910s, attempts at efficient production were under way at many firms, and scientific management elicited the same kind of evangelism it had in other industries from journalists and managers. A 1913 profile entitled "Studio Efficiency" ardently detailed the application of efficiency practices to motion pictures by early scientific producer-manager Wilbert Melville.[66] Dissenting opinions on efficiency were voiced more vaguely, usually in editorials. For example, in August 1915, *Moving Picture World's* editors expressed doubts about the growing practice, stating, "The efficiency expert who is turned loose in the studio MUST BE A MOTION PICTURE MAN," or "he will play more hob with the product of the studio than an intoxicated camera man" [all-caps theirs].[67] However, the frequent, glowing profiles of new "Efficiency Men" dwarfed these opinions.

Efficiency experts were equally preoccupied with the challenges to scientific management represented by the unique, creative, individuated nature of motion picture production. This was not an admission of defeat, by any means; on the contrary, the conflict between art and commerce was often cited as another argument *for* efficiency. Managers protested (a bit too much) that they sought not to systematize creativity, but as F. M. Taylor explained, to systematize everything else in order to "let the directors and actors work when the mood is upon them."[68] For Wilbert Melville, efficiency increased creative freedom by removing those particulars that are "ordinarily a handicap against the highest artistic results." [69] H. O. Davis, Universal's systems man, echoed this sentiment, saying that the director, freed of earthbound concerns by efficiency, was "able to think of his story, to dream of it."[70] However, practically in the same breath, these experts expressed desire for the creative control they claimed not to want by concluding that "the goal is perfect pictures," for which efficiency was necessary because "no man can produce them by himself."[71] Such insistence on the idea that perfection was possible highlights the level of managerial control efficiency sought, and also suggests its inherent tension with individual directors' creative freedom because, to increase their control over the creative product, scientific managers would progressively attempt to manage the creative process by indirect means through the management of paperwork.

Thomas Ince achieved much of his efficiency by circulating what workers on specific productions would come to know as the continuity script, to be used as a blueprint for the film, dividing tasks among departments. In addition to a synopsis and a script with numbered scenes, continuity scripts included lists of intertitles, locations, characters, and the actors playing them, as well as a cover page that featured the names of the film's writer, director, and the dates it was shot, shipped, and released. These materials, in other words, constituted a paper record of "the entire production process for efficiency and waste control."[72] Wilbert Melville innovated at Lubin by creating an editorial department where scripts were prepared "in such shape that they can be produced as written." He also relocated buildings and departments related to production so that they were in close proximity to one another, and imported new systems of accounting that kept data "segregated for each picture," so that it was possible at any time to know what various productions cost.[73] H. O. Davis took similar measures in his reorganization of Universal, where cost was estimated before a scene was taken, with the end result producing "a variation of but 5 percent from the first estimate of cost."[74] Similar planning and record-keeping systems were adopted by other companies to streamline operations, as were more rigid hierarchies, headed up by manager-producers like Melville.[75] Davis threw the entire production workforce into the same stock company, "breaking up the system of each director having his own group of players."[76]

In October 1915, E. D. Horkheimer of the Balboa Company wrote in *Moving Picture World* about his methods of regimented studio management, which included keeping snapshot records of locations, stills of every set built, a card index of props and set dressings, and a system ensuring that all of these assets, plus the less predictable human ones, were in place, ready to shoot the minute conditions were right. A former engineer, Horkheimer grounded his discussion with the disclaimer that much of pictures could not be rationalized "following a uniform method of production as the carshops do" because "artistic considerations must be allowed for."[77]

In spite of all his meticulous attentions, such efficiency sought to rationalize a process that involved—indeed, required—emotion and sentiment (for example, in interpersonal collaboration, storytelling), as well as individual judgment and subjective interpretation (in artistic and creative sectors of production, as well as acquisition of properties). Interestingly, as chapter 4 details, many of these unrationalizable qualities were also unofficial job requirements for the female secretaries who would become so vital under efficiency. The very word *efficiency* started to accrue negative connotations in the late teens as owners and managers of studios continued to seek control over an inherently uncontrollable creative process by instead controlling everything around this process.[78] Many built entire, self-contained worlds to keep assets, workers, and their practices under the highest degree of control, predicting the theme park simulacra many studios would eventually become.

Mapping Efficiency's Interests: Spatialization and Feminization (1915–mid-1920s)

Managerial oversight took fuller and more comprehensive form as it began to be designed into purpose-built spaces of production in the 1910s and early 1920s. At least that was what studio managers claimed outwardly in trade and fan magazines, which printed descriptions of studios so numerous and similar to one another that common tropes can be identified among them.[79] Discourse around the growing studios served the needs of the magazines, filling their pages with the sort of "sneak peeks" into the movie factories of the future that would interest their readership. Perhaps even more, these barely disguised press releases on companies' latest additions to their growing production plants served the studios, functioning as promotional material on a number of levels. Some self-promotion was to be expected from film companies, but the aggressive self-confidence conveyed by the tone, style, and volume of this self-promotion bespeaks a need on the part of studios to demonstrate—to themselves, each other, and their potential investors—their sense of ownership and mastery over the production process.[80]

To exhibitors, studios promoted themselves as growing, capable suppliers of enough new and diverse content to meet audience demand. To their competitors, they showed themselves as competitive and hardy, while to prospective Wall Street investors, the self-representation was meant to show them as professionalized businesses with efficient practices. Though nobody knew precisely where the business was going, everyone pretended they did by confidently appropriating strategies of industries that *were* established, incorporating efficiency methods and buying up as much of the supply chain and means of production as possible—aspirational texts in which managers and owners fantasized total control.

All of this self-assuredness would have lasting consequences for the women whose success as important players in the early industry was possible precisely because the fledgling business had little fixed sense of self, or rigid notions of what (or more important, who) it should look like. Those circumstances began to change as trades circulated mapping schema, representations of studios' interests, embodying their authors' biases and values through presences (what was represented or emphasized) and absences (what was minimized, marginalized, or left out). The mapping schema showed the interests of efficiency itself (order, hierarchy, control, standardization) and how it was applied to everything and everyone at studios.

Early film studios were typically located in existing houses and mansions that were adapted for their purposes.[81] However, in the mid-1910s, larger studios began to be built specifically for the purpose of moviemaking for greater production capacity. Published layouts and descriptions from the 1910s and early 1920s document the creation of new departments both to specialize in and anticipate the various needs of expanded production, and their location near related departments. Layouts are commonly explained through straight description of buildings by size, shape, purpose, and spatial relationship to other buildings and sectors of production. Eugene Dengler's 1911 description of Selig's "Diamond-S" Plant in Chicago started mapping the space from the road ("You notice somewhere the sign 'Selig Polyscope Company'") before continuing into the main building and indicating the layout and contents of each floor complete with square footage

("On the third floor is the studio proper, an enormous room, 179 by 80 feet, whose solid glass walls and roof rise two and a half stories above the floor") and such special features as air conditioning systems and elevator size.[82]

Later, many such descriptions were published of West Coast studios, so detailed and specific as to the spatial relationships that one could draw a more or less accurate map from them.[83] The building of Universal City was detailed at different stages in several trade articles in 1914. One example described each building on the lot as representing "a different kind of usefulness," with edifices purpose-built so their interiors served one function or department (for example, an administration building, employment agency, and hospitals and infirmaries with onsite doctors and nurses), while exteriors doubled (and sometimes tripled and quadrupled) as film locations, with different facades on different sides of the same building.[84]

Motion picture assets—from props and cars to exteriors and—to some extent—human beings—were increasingly acquired and stored as efficiency sought to convert all of production into a matter of asset and resource management. However, an inefficient sense of relish characterizes published descriptions of the wealth of assets at various companies written in a tone not unlike the newsreel from the opening of *Citizen Kane*, with its exclamations over the array of exotic animals, buildings, art, and so forth in Charles Foster Kane's Xanadu. The lists of assets communicated bounty, celebrating the studio-as-warehouse, where necessary tools for any eventuality had been acquired, organized, and stored. One report began, "There is rapidly rising a city capable of accommodating 15,000 souls, built for the express purpose of making motion pictures," and then arrayed Universal's bounty, or the potential for it, before the reader. The wardrobe department, it explained,

> contains wardrobe of every conceivable sort, which is valued in the rough at about $35,000. In addition to this the costume shops which are nearby are so arranged that they are built to turn out the designs which are covered by every period of dress from the era of palm-leaf girdles to the present time. Its costume shop contains twenty electrically operated sewing machines and the work is supervised by one of the best costume designers obtainable, who is able to outfit a picture with the proper costumes demanded by any age.[85]

Here, the work itself (how the costumes are designed and whether they'll make movies better) is seemingly irrelevant, consigned to the shadow of the real challenge—acquiring the right machines, workers, and supervisors—that had already been met.

Thomas Ince's Inceville and others were profiled through lists with a similar sense of luxuriousness about them, delighting in the plenitude of studio resources they represent.[86] In August 1915, it was even reported that Ince, "determined to have a studio equipment absolutely independent of weather conditions," had eliminated the necessity of the pesky sun altogether via the installation of "a new big power plant at Inceville" that came with its own list: "two three-cylinder engines, each of 125 horsepower, two thirty-five-kilowatt direct-connected Westinghouse generators, and a four-panel marble switch board, together with all the numerous smaller accessories."[87] Not to be outdone by Ince's private sun, Paramount would later purchase its own forest. The tract of timber land was described in a 1918 issue of *Photoplay* via a list of the assets used to control lumber through every stage of life, including

> a private sawmill and steamers to carry the material to San Pedro harbor. The carpenter shop, planning mill, and other equipment for making the timber into scenery are organized on a scientifically economical basis, so that no splinter is wasted. Every stick is taken care of, after a set has been dismantled; sawdust and shavings are stored for various uses. And when a piece of wood has done its bit, it goes to the furnace room and serves its purpose even in death.[88]

Assets were also frequently arrayed in trades through photos with captions conveying variety, vastness, and bounty. *Motography* accompanied its ten-page article on the Lubin "Diamond-S" Plant with multiple lists of assets, as well as photo lineups of its well-known actors and unusual assets, including Todders the elephant, "the biggest actor" at the studio.[89] Such photo arrays appeared regularly in the magazine by 1915, under titles like "Some Views of Metro's Gigantic New Studio," emphasizing scale and bounty through such captions as "A partial view of the big studio, which does not even faintly indicate its immensity."[90]

Maps and mapping schema such as diagrams and panoramas were similarly drawn and discussed in terms of control and bounty.[91] A June 1919 *Scientific American* feature on William Fox's plans for a new studio in New York was illustrated with a map of the three-story factory that promised to solve the problem of movie companies' various parts being "scattered through city suburbs and far out in the country—wherever land and space could be obtained," with "the executive offices downtown, the laboratories uptown, a workshop in the business district, a storehouse in the factory district, and a property room in any available spot."[92] The Fox studio was mapped via a 3-D drawing of the building, with sections removed to show their contents. Again, a sense of vastness is conveyed in descriptions of departments and the studio floor "where 100 scenes may be enacted at one time."[93] A 1918 *Photoplay* article entitled "A Bird's-Eye View of the Lasky

Waiting for Their Cues.   "Old Hickory," the $2,000 Coach.   Working Three Sets in the Studio.
Looking Across the Yard.   "Toddles," the Biggest Actor.   The Studio, Viewed from the Yard.

Fig. 4. Asset Array: "The Wonders of the Diamond-S Plant." (*Motography*, July 1911.)

Studio" not only listed assets, but also mapped out their efficient segregation from one another. The map displayed the newly expanded studio's
buildings grouped according to function, declaring the layout an improve-
ment over previous configurations in which "in the one building, actors
were engaged, accounts kept, scenarios written and scenery built." The old
multipurpose building (marked 11 on the map in figure 6) was now just the

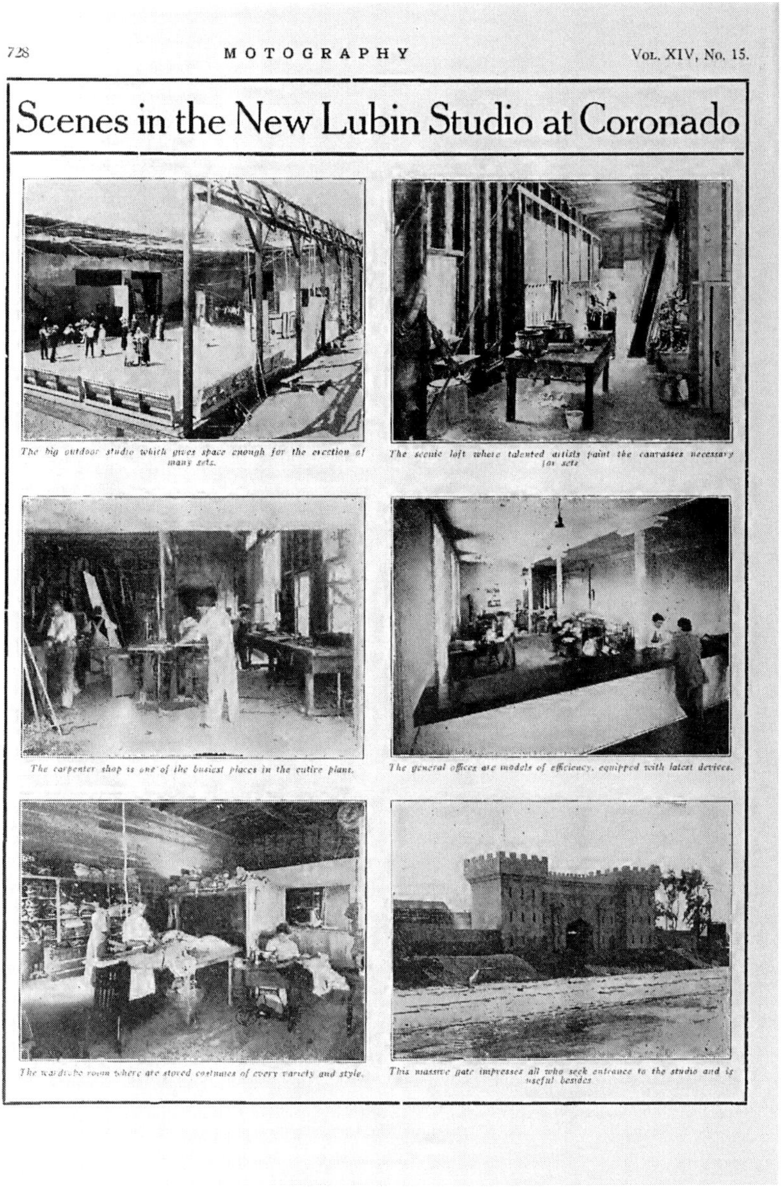

728   MOTOGRAPHY   Vol. XIV, No. 15.

Scenes in the New Lubin Studio at Coronado

The big outdoor studio which gives space enough for the erection of many sets.

The scenic loft where talented artists paint the canvasses necessary for sets

The carpenter shop is one of the busiest places in the entire plant.

The general offices are models of efficiency, equipped with latest devices.

The wardrobe room where are stored costumes of every variety and style.

This massive gate impresses all who seek entrance to the studio and is useful besides

Fig. 5. Photo Spread: "Scenes in the Lubin Studio at Coronado." (*Motography*, October 1915.)

property receiving room, in a plant that covered two city blocks and was a paradise of assets controlled down to the inch by the fire and safety departments which patrolled the studio day and night.[94]

Such materials asserted studios' autonomy and sovereignty over human resources as well, typically listing and mapping workers along with other assets.

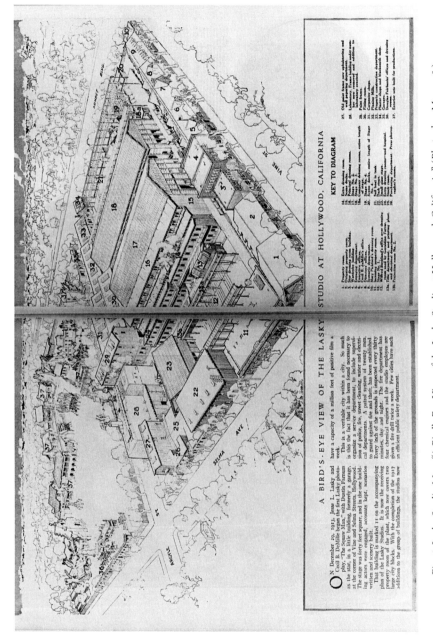

Fig. 6. Lasky Studio Map: "A Bird's-Eye View of the Lasky Studio at Hollywood, California." (*Photoplay*, May 1918.)

Some studios made even more explicit attempts to mark ownership and code labor. Balboa marked its property by painting buildings green with white trim, and carried the color scheme through the whole plant down to its correspondence, which was written in green ink on white stationery, then "mailed with the green one-cent Balboa exposition stamp."[95] Used as a point of interest around which to promote the studio in publications like *Motography*, this uniform color scheme served as an early form of integrated branding. World Pictures Studio color-coded members of its crew ("Stage Hands, blue. Property men and carpenters, white. Electricians, brown") to eliminate the confusion that had previously arisen "when a director wanted a piece of furniture or prop removed from a set" and didn't know which crew members were property men ("All he has to do now is look for a man in white").[96] Employees at Famous Players were assigned color-coded badges that were later abandoned when they didn't show up under the lights on set.[97] Though many such measures proved unsuccessful, these measures, as reported by the trades and fan magazines, were nonetheless effective in communicating the studios' message that they were systematized and controlled from top to bottom.

The new studios divided production among units headed by directors, and later producers.[98] Production labor became increasingly specialized, as jobs were separated to simplify and standardize their tasks, just as the departments had been separated geographically. The studio manager, who sat atop this hierarchy, rose even further in terms of status, a shift reflected by numerous profiles of efficient studio managers around 1915. A profile of Milton E. Hoffman, studio manager at Famous Players–Lasky in 1918, states that "with the increasing activities" of his studio, "the position of studio manager has become of prime importance."[99] The sentiment is echoed in profiles of Melville, Horkheimer (both E. D. and his brother H. M.), and H. O. Davis.[100] Though they emphasized the minimal impact of systematization on the work of the director (other than to free him to focus only on the creative process), profiles could not hide that the creative process, along with every other link in the chain of film production, had changed in that all of it was now routed through or relocated completely to a manager's office.

The specific impact of such changes on women's labor and professional identities within film production begins to emerge in the following description of manager Milton Hoffman's workday:

> Arriving at his desk early in the morning, Hoffman disposes of the mass of letters and telegrams and telephone calls and makes a round of the studio, and confers with the chiefs. His travel through the studio is usually stopped every ten feet by someone with a query, but he goes

through the day unruffled. He has systematized his organization so as to obtain a maximum of efficiency with a minimum of friction, but no man regards him as a "boss"—he is and prefers to be considered "one of the boys."[101]

Though Hoffman professed himself to be "one of the boys," he probably spent much time in the company of women, given the clerical labor required to dispose of masses of letters, "confer with the chiefs," link separated departments, and supervise every aspect of film production from a central hub. Increased paperwork was the only way to efficiently reroute the production process through such a rigid, hierarchical chain of command, and shift decision-making and resource management responsibilities from individual creative workers on separate productions to executive-managers without creating more of the dreaded "time wastage" that scientific managers deplored.

A Place for Every Girl and Every Girl in Her Place: (Re)Locating Women through and on Paper (1910s–early 1920s)

Studio managers made few direct references in published interviews to the increases in paperwork generated by their cost accounting, indexing, record keeping, centralization, and other efficiency measures, and mentioned growth in clerical staffs even less. Similarly, trade reports seldom devoted space to clerical hirings and referred only occasionally to clerical workers, as in a brief report on the Edison Studio's reorganization that discussed moving the office staff (including a female aide) to new offices at the studio to "centralize each department's work for greater efficiency."[102] More often, these workers were noted indirectly, or their presence was implied by descriptions of studio improvements.

And yet, clerical expansion was most certainly happening all over the film industry, not just at those companies that disclosed it to trade reporters, not just in business offices where films were sold and financed, but on studio lots where clerical staff were incorporated into film production. As it had in other industries, scientific management created efficiency through increased paperwork, necessitating larger clerical staffs, the cost of which was controlled by employing low-wage female underlings in new clerical roles. Thus, even when the clerical workers, male or female, are not specifically named in trade journal discussions between studio managers and reporters, it is nonetheless possible to track the clerical expansion at studios by tracking the paperwork efficiency created. This paper trail suggests how and when certain types of work related to filmmaking became understood as women's work, and how related gender-normative assumptions guided women's subsequent participation in film production.

Each aspect of comprehensive managerial control enacted at studios expanded the various forms of planning, accounting, record keeping, and message circulation. This expansion occurred first at business offices of studios, where global cost and process control were initially attempted. For example, in 1915, the V.L.S.E. Company announced that it had partitioned its New York offices to separate each of the company's executive branches, including publicity and accounting. The *Motography* photo spread of the new offices, each of which now included desks and chairs for multiple under-lings and modern equipment for clerical and record-keeping processes (file cabinets, inboxes, typewriters, stenographs, and so on), showcases staffers in the accountant's office in their proper sex-segregated roles, with men at the accountants' desks and a woman at a typewriter.[103]

Based on the rationales offered by managers, many expansions were made not simply to keep pace with a growing business, but to increase efficiency via a more formal hierarchy, better record keeping, and cost accounting. Formalization of hierarchy at one company was reported as "centralizing the authority in the home office" to replace an older system

The accountant's room.

Fig. 7. The Accountant's Room: "The Executive Offices of V.L.S.E." (*Motography*, July 1915.)

in which authority was more evenly distributed among managers of its various branches, and appointing efficiency experts to standardize best practices (from shipping practices to "the appearance of an office") across all branches.[104] Similarly, when Metro Pictures Corporation moved its New York offices "to provide room for an increase in the executive and clerical force," it centralized various planning and executive branches, previously on separate floors, in one suite, where each department had a separate office, with clerical and planning departments (such as publicity and mailing) adjoining one another on one side and executive offices grouped together on another.[105] In 1913, a visitor from Universal's West Coast Studios wrote at length in the studio's house organ of the impressive operations taking place largely on paper at the company's New York headquarters. In a series of features, the visitor described the office's ten trunk phone lines and fifty extensions (handled by Miss Lillian Clair McGuinness), its "one-hundred percent" efficient accounting department, stenographic department with a "galaxy of pretty girls" operating their machines at astounding speed, a complex scenario processing and purchasing system, a payroll system providing for four hundred East and two thousand West Coast employees, and three "boy" clerks filing three thousand pieces of paperwork each day.[106]

Paper-regulated efficiency was soon adapted to the process of film production itself through the centralization of clerical and paper-based departments into one production office, overseen by the general manager, that served as an informational clearinghouse for all departments.[107] Though buildings were seldom named for the paper workers who resided there, the increased importance of written planning was made plain by its place in the published layouts of studios, which included buildings or floors specifically marked out for executives and administration. In 1913, the Thanhouser studio in New Rochelle relocated its administrative staff from the "factory end" of the plant to a new, purpose-built structure on the other side with "special rooms for the bookkeepers and stenographers," as well as "uniformed attendants" stationed at the gates.[108] At Lubin's Coronado plant, offices and administration sat at one end of the lot, with production facilities at the other. Centralization of management was engineered into the new Fox studio building "by authorities on factory and office construction, with a view to speed, economy, and concentration in every possible phase of efficient motion-picture production, from filming to bookkeeping to stenography to starring." The map of the Fox building duly depicts the administrative offices "from which the activities of 63 branch offices covering the world are directed" as centralized on the lower floors at the front of the studio.[109]

Though the clerical buildup was only beginning in the late 1910s, its importance to production could already be seen in the 1918 Lasky map

featured earlier in this chapter. In figure 8, the Lasky map has been modified to illustrate this discussion, with shading to indicate sites of primarily
paper-based processes. Buildings used for planning, record keeping, and
cost-accounting were centralized on the east (Front) side of the lot with the
"5. Engaging department" (early employment office), "6. Executive offices,
7. Cecil B. deMille's office, 8. Directors' offices, 9. Scenario department," but
also included a purchasing department on the other side of the lot nearer
the craft departments for onsite cost accounting.[110] The map does not refer
to the secretaries, stenographers, and clerks who would also have occupied these buildings, but the location and existence of such paper planning
departments indicates that Lasky's is no longer a production system involving only occasional paperwork by a purchasing clerk here and a private
secretary there.

At the same time, the literal marginalization of clerical work on the map
indicates female workers' exclusion from creative and technical production
specialties in a system that would increasingly associate them with typing,
filing, and answering phones. Wilbert Melville uses a domestic metaphor
when describing his plant's efficient layout, saying that

> the relative locations of buildings were determined with a view to the
> part they play in the general scheme. Those which are most used in
> connection with the work on the stage are grouped around the stage.
> To use a homely illustration, when a producer goes into the studio
> to make a scene he finds everything as conveniently placed for him
> as does a housekeeper in a well-planned kitchen. There is no time
> or effort wasted because the property room, for example, is located
> remote from the stage.[111]

Under the developing informal system of sex segregation that Melville's
convenient rearrangement entailed, a large portion of female workers at
studios would now be nowhere near his metaphorical "kitchen." On the
Lasky map, a similar spatial separation can be observed between production and the management sectors clerical workers serviced, with various
clerical buildings located together at the front of the studio near other
paper processes. As figure 8 indicates, clerical workers were farther from
the stage and related production sectors than they had been before, and
thus farther from production jobs. At other studios, clerical and production processes were set apart in much the same way. At Universal, the
administration building was located on the western edge of the property,
with laboratory, purchasing, and city buildings such as the barbershop
and drugstore, and production buildings were located at the east end,
separated by a service road.[112] Aside from female actors (Mary Pickford

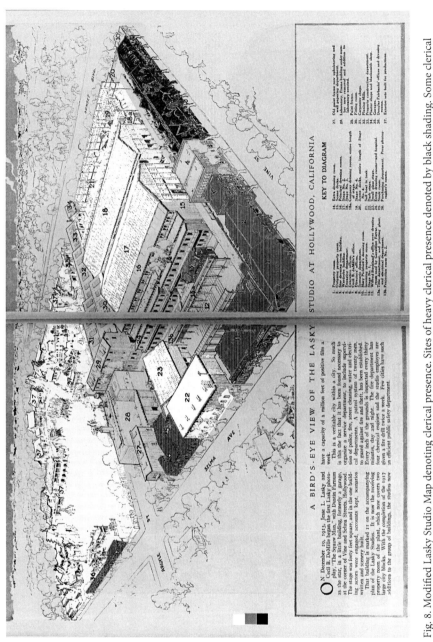

Fig. 8. Modified Lasky Studio Map denoting clerical presence. Sites of heavy clerical presence denoted by black shading. Some clerical presence denoted by light gray shading. Spaces where films were shot marked with white shading. (Map modified by author from original in "A Bird's-Eye View of the Lasky Studio at Hollywood, California," *Photoplay*, May 1918.)

and Geraldine Farrar), the Lasky map neither described nor implied the presence of major female creative figures. Studio founder Jesse L. Lasky, along with male creative figures Cecil B. DeMille and art director Wilfred Buckland, are all mentioned by name, while female directors, producers, and executives are largely absent from the descriptions, photo spreads, and other reports on new studios.

Writing of imperialism, sociologist John Galtung notes the discrepancy between a culture's center and periphery, delineated by the politically dominant people or nations, as a form of structural violence.[113] Delicately adapted, these concepts aid an understanding of how studios' systemic shifts affected women's prospects. As studios worked to project an image of professionalism and control, they relegated women to more culturally acceptable roles. Increasingly, women were represented in such discourse not as individual creative or managerial figures, but as groups of "girls" who worked behind the scenes under male leaders. However influential or important some women continued to be, women in general had been rooted to a specific, peripheral place at studios, literally and figuratively. This structural narrowing of women's prospects took shape after studio managers began to think about how they looked to the outside world.

Paper Efficiency: The Enlarged Role of Clerical Processes in Studio Production Systems (1910s–1930s)

With administration buildings and central offices as clerical hubs, work at studios was organized and managed through the record-keeping and filing systems and the ledgers mentioned by E. D. Horkheimer and his peers. As larger plants opened on the West Coast, onsite accounting was ramping up to match that of more impressive operations back east. Less than a year after the western visitor marveled over the efficiency of the New York office paper pushers and number crunchers, an eastern visitor for the same house organ reported "some thirty bookkeepers scratching in unison," at the West Coast studio, where there was now "not a single picture made" whose cost couldn't be accounted for by the penny.[114]

Work in production departments was also organized with new indexing and filing systems, frequently discussed in managers' reports on studio organization. One report from 1915 states, "Each item in our property rooms—and there are a hundred thousand of them—is card-indexed so that it can be found on the instant. All props must be kept clean and dusted. We have a complete stock of furniture of all periods and rent nothing."[115] In 1922, Lasky manager Melvin Riddle described the fully systematized props department as having two files, one with the 65,000 props the studio owned and one with information about many more items and where they might be obtained.[116] In another article, Riddle explains that every costume "is systematically indexed and listed and the members of the department can lay their hands on any certain costume or article at any time within a moment's notice."[117] In addition to stockpiling exteriors on their lots, as in the case of Universal, studios began snapping and indexing pictures of off-lot locations and "every set erected on stage" in a "location book," to assist producers in finding "the spot they need without unnecessary loss of time."[118]

Newly differentiated planning departments also sought a global form of control through record keeping and abundant paperwork. Chapter 5 expands further on the shift from stock companies to a different kind of casting (known as "typage") through locating and putting talent under contract. This process was already under way by 1915, when it was reported that Lubin would "do away with the special company system," in which each director had his own group of actors, in favor of a single, large company from which casts could be planned for the entire studio.[119] The same year, Premier hired a theatrical agent "to carefully tabulate all valuable data with regard to types suitable for use in motion pictures."[120] Several years later, Riddle reported that an "average casting office" had "a very complete set of files which are cross indexed to save time and make them more practicable, with large cards for principal, freelance and extra players," listing "his or her height, weight, and other physical data."[121] Increasingly, selling films would involve selling stars, as well as managing their relationship to the public.[122] At the same time as it appointed its new scenario and casting department heads, Premier announced E. A. Levy as head of the publicity department, the third part of the process by which products were planned for eventual sale to the public.[123] In 1917, Essanay announced a new "investigation department" to determine what its sales organization, theater owners, and public wanted by tracking feedback from business partners, but also "to keep a record of critic's reviews, and secure all available information from the public," all for use in selecting subsequent projects and stars.[124] Processing fan mail, as well as generating interest and revenue through fan magazines' articles, ads, and contests, was becoming part of an increasingly systemized, paper-based process around publicizing films, stars, and celebrity.

Perhaps the biggest increase in paper and paperworkers came in those parts of the preproduction process related to the script. In the name of efficiency, Ince's use of screenplays as films' blueprints was widely adopted in the early 1910s, and studios circulated continuity scripts to various departments to serve as their marching orders. This meant more copies of scripts, to be sure, but that was just the last step in a larger process—before scripts could be circulated at all, they had to be produced through other paper-based procedures. Where once directors or cameramen had supplied the plots for pictures, the increased volume of production as well as the increased sophistication of the viewer necessitated more and different stories, which studios acquired through a number of means. One of the most important of these was to solicit scripts from writers outside of the studios, both professional and amateur. Mail-in solicitations alone no doubt greatly increased studios' clerical needs. The number of submissions accepted for production was estimated at 1 percent in 1911, implying more than a few sets of eyes required for the task of weeding out those 99 out of 100 submitted scenarios that would not suit.[125]

A 1913 *Motography* feature by Mabel Condon detailing what happened to scenarios once they arrived at film companies (and Essenay in particular) portends the variety of workers that the process would eventually necessitate:

> On receipt of a scenario its prompt acknowledgment is made by the scenario editor in the way of the printed postcard with the name and reference number on the manuscript inserted. . . . The story is read by the scenario editor and, if rejected, is returned to the author with the enclosure indicated in form 3, the reason for its return being checked off with a pen or pencil mark as shown in the example.[126]

The writer goes on to describe the fate of a more successful manuscript, detailing the process of the editor's "approval of it on a blank for that purpose, using a carbon sheet" before it was sent to a producer, who signed off it, then returned the original to the scenario department, and kept a duplicate for himself. For the same script, approval was also sent to the writer on a form created for the purpose, with a copyright blank, and so on. Yet more forms were dedicated to the script's preparation for production. The director dictated "his version of the story to his stenographer," assigned a cast from his company, compiled a list of props, and diagrammed each setting on another form created for the purpose. All of this paper was used in shooting the script, and when physical production was completed, a producer's copy was "typed on crisp white paper and the whole is given a backing of heavy blue paper which will be filed for possible future reference."[127]

In addition to the script itself, six forms were used in the submission and planning process as described by Condon. Clerical staff involved in this process would have included some or all of the following: (1) the worker(s) who logged in the scripts, assigned them reference numbers, and sent correspondence to rejected authors; (2) the worker(s) who generated the forms and filed and recorded them; (3) the stenographer who typed the director's dictated script; and (4) the worker(s) who created the finished "producer's script" document and filed it. Some of these clerical roles might have been

What Happens to the Scenario
By Mabel Condon

EACH motion picture producing company probably has its own individual method of dealing with the manuscripts submitted for its approval. In dealing with the fortunes and misfortunes of the scenario, from the brain of the author to the film synopsis, we will instance the method employed by the Essanay company.

The way of the scenario is devious. If it is accepted the incidents which center around it thereafter are many and varied, and if it is rejected it must try, try again.

On the receipt of a scenario its prompt acknowledgement is made by the scenario editor in the way of a printed postcard with the name and reference number of the manuscript inserted. It reads as shown in form 1.

The story is read by the scenario editor and, if rejected, is returned to the author with the enclosure indicated in form 2, the reason for its return being checked off with a pen or pencil mark, as shown in the example.

If the reason for the manuscript's rejection comes after the eighth reason listed, a postcard giving information, as in the following, is also enclosed:

Arrange your story in scenario form. A synopsis of about 200 words followed by short scenes. All manuscripts must be typewritten. We are in the market for original dramatic stories with strong strong heart interest for short stories with unusual themes and for bright sparkling high class comedies. We are not soliciting Western scenarios, costume plays, war stories or plays with foreign settings. Our prices vary according to the merit of the story. Address all manuscripts submitted, to the ESSANAY FILM MFG. COMPANY, c/o Scenario Department, 1333 Argyle St., CHICAGO, ILLINOIS.

And, as far as the company is concerned, that is the end of that scenario, unless it is rewritten to make the kind of a story for which the company happens to be in the market.

But, should the scenario be available, its tale is a happier one. That of "The Gum Man" is herewith instanced.

The story, as the scenario editor received it, was neatly typed on five sheets of Robin blue paper, typewriter size and from the general appearance of the copy, it was evidentally the work of someone who had given scenarios and their making careful and valuable thought. At the foot of the first page was the information, "An extra carbon copy of this script will be supplied to purchaser upon request."

This is how the scenario read. (It will be noticed that the author's title was changed):

BY GUM!
Rural Comedy. Twenty-four Scenes. Four Exterior, Four Interior Settings Required.
SYNOPSIS.
Fred Smith, chewing gum salesman, stops over at Cobb's

Corners to introduce his wares at the general store. A Trubbell Hunter, constable, has just "got in" with a Detective Bureau and is highly elated. He has several run-ins with Fred and becomes antagonistic. Fred meets Mamie, hotel waitress and chambermaid. They like each other. Hunter gets word that a notorious burglar, "Iron-jaw" Pete, is thought to be in the neighborhood. This criminal's distinguishing trait is a fondness for gum-chewing. Fred is suspected. Everything points straight to him, in Hunter's estimation. Next morning a robbery has been committed, a valuable necklace is gone. Fred, ignorant of this, buys a cheap necklace to present to Mamie. Hunter surreptitiously sees the presentation, hastens away for a warrant and hurries back. Meanwhile a friendly hotel clerk has "tipped off"

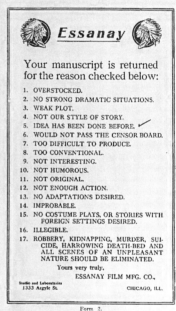

Essanay

Your manuscript is returned for the reason checked below:

1. OVERSTOCKED.
2. NO STRONG DRAMATIC SITUATIONS.
3. WEAK PLOT.
4. NOT OUR STYLE OF STORY.
5. IDEA HAS BEEN DONE BEFORE. ✓
6. WOULD NOT PASS THE CENSOR BOARD.
7. TOO DIFFICULT TO PRODUCE.
8. TOO CONVENTIONAL.
9. NOT INTERESTING.
10. NOT HUMOROUS.
11. NOT ORIGINAL.
12. NOT ENOUGH ACTION.
13. NO ADAPTATIONS DESIRED.
14. IMPROBABLE.
15. NO COSTUME PLAYS, OR STORIES WITH FOREIGN SETTINGS DESIRED.
16. ILLEGIBLE.
17. ROBBERY, KIDNAPPING, MURDER, SUICIDE, HARROWING DEATH-BED AND ALL SCENES OF AN UNPLEASANT NATURE SHOULD BE ELIMINATED.

Yours very truly,
ESSANAY FILM MFG. CO.,
Studio and Laboratories
1333 Argyle St.   CHICAGO, ILL.

Form 2.

Fred, who was easily convinced the clerk of his innocence. Fred proposes a hasty marriage to Mamie; she agrees. As they are ready to start Hunter gets back. In order to get away from him Mamie devises a great scheme. It succeeds and they rush off, leaving Hunter stuck fast BY GUM to a chair. They rush to the station. Hunter, after herculean struggles, drags himself from the chair but the seat comes with him; in this position he hastens to the station in time to grab Mamie and Fred. But the real crook, who had ambled through the story unostentatiously, is now discovered by a trick of fate, and all ends happily for the lovers and embarrassingly for the over-zealous and misguided Trubbell Hunter.

CAST.
A. Trubbell Hunter .... "Rube" detective-constable
Fred Smith .... A young chewing-gum salesman
Mamie .... Hotel waitress and chambermaid

Form 1. (image content)

CHICAGO, ILL. Feb 6 191_

We have received from you today the following manuscript:

Dog Gone Dog

which will be given careful consideration. Writers should retain carbon copies of any scenarios submitted to us, as we are not responsible for MSS. lost in the mail. Sufficient postage for return of MSS. must accompany all contributions if it is desired they be returned.

Yours truly,
Editor of Scenarios,
ESSANAY FILM MFG. CO.
1315-1333 ARGYLE STREET

Should you inquire about the above scenarios state that reference number.   Ref. No. 763

Fig. 9. Scenario Forms: "What Happens to the Scenario." (*Motography*, March 1, 1913.)

performed by editorial staff, and one worker might have combined several roles, but given the expansion studios would undergo, the number of such positions would likely have multiplied. By 1919, one scenario editor wrote that, with few exceptions, "the editorial department of a picture concern has an outer room where a staff of young women opens and files incoming mail and correspondence."[128]

Still more workers were needed to prepare scripts for production. Previously, directors had chosen or written their own scripts without the involvement of the script or scenario department. The process at Lubin was reorganized in 1913 "along more modern lines," requiring collaboration with the script department, which would now serve the dual purposes of "handling the incoming scripts as has always been done," and putting those scripts approved for production "in perfect technical shape before they are handed to the director for production."[129] And where previously "the director would have revised and edited a script for production, the task would now be handled by trained staff writers."[130] In this way, the Lubin scenario department would prepare "perfect working scripts" to be cast and approved before being "given a director who will make the production without departing from the lines laid down, though any suggestion from the director will be welcomed by the script department."[131] In 1917, Fox's scenario director wrote that his department would not only write scripts but also, using a similar system, shepherd projects through production and exhibition, making "for an enormously increased efficiency in the production of perfect picture stories—perfect pictures and perfect stories—a harmonious fusion of the literary and the mechanical."[132] The title of "continuity writer" was assigned to the job of writing detailed shooting scripts of accepted scenarios that contained "all that is embodied in a production before thousands of dollars necessary to make it are appropriate."[133]

DeMille screenwriter Jeanie MacPherson gave a more frank assessment of the role continuity writing played in managing the creative process, calling it a means to "director-proof" a script by making it "so complete and perfect in all its details that no one but a fool could go wrong in its direction." However, she added that a skilled director "WANTS the complete, detailed script," because "his work is made easier by having all the details before him, to consider and use unless he has something better to substitute."[134] In 1917, Universal's H. O. Davis declared that in addition to being devised by writers in all their various forms (scenarists, continuity writers, dialogue writers, originators of adapted plays, books, and so on), at least three other supervisors (scenario, editorial, studio manager) would check over and adjust each script selected for production before it was given to the director to make his pass through the material.[135]

"Readers"—members of the scenario department who read and reported on submissions and who were often female—assumed a position between clerical and editorial/writing sectors in their role of filtering incoming material for producers, directors, and executives. Readers began to be mentioned more frequently in the 1910s in trade discussions of screenplay acquisition and were soon an integral part of the process at many firms, as demonstrated by a 1917 *Photoplay* column that cited reading as a means to break into screenwriting, reporting that a great many were "being employed by the various companies to assist the scenario editors."[136]

Of course, spec script submissions were not the only sources of material. Plays, books, and other published material were considered as well, whether they had been submitted or not. Universal's readers, H. O. Davis mentioned, "systematically read all the fiction in the current magazines." Davis also noted that after readers processed material from all sources, a short synopsis and opinion was attached, and then passed on to a scenario editor, who culled 95 percent. The remaining 5 percent of submissions were sent to a second reader and the process was repeated before the remainder was sent by the editor, "with his recommendations for purchase, to the production manager, who has a reading staff of his own, trained to read manuscripts not only from the story standpoint, but from the production standpoint."[137] In 1921, Bradley King, a female member of the Ince scenario department, observed that all large studios had readers, any of whom might arrive at work each day to find their desks piled high with "seventy original scenarios, besides a score of magazines and books, all shouting to be read—and all one day's offerings!" King went on to describe the reader's job, including its more clerical aspects, saying, "We open all the envelopes, get ready our little cards for criticism and filing—then get busy, reading, writing criticism on rejection cards and pass anything worthwhile up to an editor."[138] In 1922, Kate Corbaley described an independent story service that had further systematized the process of sifting through published material by sending "daily to every studio, synopses of all novels and magazines published in America, as stage plays produced in English, and all worthy original scenarios."[139] The service did not become an industry institution, most probably because it did not fit with the studios' interests in gathering labor (in this case, the readers) under one roof along with the written material they acquired.

Eventually, as Tino Balio summarizes, "all the companies had story departments with large offices in New York, Hollywood, and Europe that systematically searched the literary marketplace and the stage" for material.[140] By the 1930s at the major studios, "Twenty-five thousand pieces of material, in the form of short stories, articles, books, typed scripts, galleys, and plays were not only read each year by studio analysts but synopsized and distributed to the studio's thirty to fifty producers," and purchased based on their interest.[141] At Fox, department head Julian Johnson issued a bulletin based on reader recommendations so that producers could read synopses and request copies of the full material as necessary.[142] Harry Warner eventually rationalized his studio's acquisition process further by using the studio's gross income to determine the story department's budget.[143]

Reference libraries, also frequently staffed by female research clerks and librarians, were built in or around scenario departments, and their contributions to the paper-based planning process were described in all their efficient glory in the mid- to late 1910s at Selig, Balboa, and Universal.[144] Eventually, these libraries grew into their own separate-but-related research departments. Lasky's research department was in operation by the end of the 1910s, and MGM's in 1924.[145] By the 1930s, all the major studios had followed suit with devoted departments that authenticated period settings, costumes, and props using "reference libraries containing art history books, prints and illustrations, and art and architecture magazines."[146] Like reading, research was ascribed increasing importance and related to efficiency by studio managers like Melvin Riddle, who proclaimed that library students visited the Lasky "fact storehouse" to study the impressive systematization of its "voluminous masses of matter, keyed by the files and indexes." The library's staff, he continued, provided details about historical pictures to scenario and design departments, and added to the studio's research collection, "continually working ahead, storing up data and preparing against possible emergencies."[147] MGM's research department, begun by production designer Cedric Gibbons, amassed "the largest of all movie studio research libraries" and held "20,000 books and 250,000 clippings cross-referenced on 80,000 index cards" in its buildings, with staff to research up to 300 queries a day.[148]

In addition to collecting, filing, and organizing materials such as "plates and books" for use by production in the manner of a typical librarian, researchers produced written and photographic background research for use in the writing of films.[149] In some instances, research helped steer the direction of a project before it was written. Interesting historical facts turned up by the department about Francis Drake motivated Howard Koch to reshape *The Sea Hawk,* which he had been assigned to write as a typical formula adventure vehicle for Errol Flynn, into a more character-driven swashbuckler with increased substance and credibility.[150] Researchers also created briefs on periods and locations for use in production. Sometimes referred to as "bibles," briefs outlined relevant information for departments.[151] Like other departments related to planning and script, the research department's chief product was clerical output, the intermediate good (research documents) that was necessary for production of the final product (finished films) to go forward. Studios' legal departments multiplied clerical output as they came to work more and more closely with writers, sometimes on a day-to-day or even page-by-page basis, to manage risk around plagiarism, copyright, and libel.[152] Still more paper was created by the censorship departments that arose with the Production Code Administration as studios attempted to anticipate code violations and eliminate them before they were flagged by the Breen office.[153]

These examples highlight the growing significance of the script within production but also—and more relevant for further discussion of women's labor identity in the studio system—the greater importance of paper generally, and clerical output specifically. Under efficiency's regime, the ranks of scenario departments swelled to include not just a departmental head and staff writers but the script-clerical hybrid positions of reader and researcher, and the strictly clerical support staff that made their work possible: typists, stenographers, clerks, and others to generate, copy, circulate, and file scripted material and messages related to it. In publicity around the buildup of the scenario and related departments in the late 1910s and early 1920s, stenographers and other clerical or hybrid positions were mentioned alongside writers with increasing regularity in descriptions of scenario department work. In "How to Write Movies," Anita Loos and John Emerson write, for example, "We find it easiest to dictate our scenarios—saves time and facilitates concentration on the work," adding "When you're under contract to turn out a new photoplay every two months you'll hire a stenographer, too."[154] Though this instance of freelance writers hiring freelance stenographers represents an ad hoc use of clerical labor rather than a systemic one, the article, which included a photograph of the female stenographer taking dictation, treats stenographers as an accepted component of the writing process. In a description of Keystone's scenario department, hiring stenographers was described as a formalized company practice:

> The offices occupied by the Keystone scenario department are being completely renovated and refurbished. Soft carpets, easy chairs, subdued colors and every little detail of comfort and restfulness have been resorted to with the idea that the staff of writers will do better in the improved surroundings. A library of reference books, individual stenographers for each writer, dictating machines and other conveniences are combined in the most up-to-date scenario department in the west.[155]

The author's somewhat disconcerting inclusion of stenographers with books and dictating machines in the list of "conveniences" hints at the developing logic under which efficient scenarios required paper and the people and machines to process them. By the time Melvin Riddle wrote about the Lasky scenario department in 1921, it was common practice for studios to have devoted story and/or scenario departments served by large staffs. In addition to its screenwriters, the Lasky scenario department's "very large personnel" included a chief supervising director, a film editor "and his assistants," as well as "Miss F. M. MacConnell, and the scenario reader, who reads and passes upon all outside contributions, several title writers, a number of scenario writers, and a host of clerical assistants, stenographers, cutters, etc., under the direction of Miss R. B. Miller, chief clerk."[156] The inclusion of clerical workers in this list of employees—necessitated by "the growing importance of the story in film production"—denotes their growing importance to the studio's efficient machine, which could not function adequately without "each cog working in perfect unison."[157]

Conclusion

Riddle's reference to a female reader and chief clerk hint at the direction in which women's labor at studios was being channeled and the workplace identity that would, more and more, narrow their prospects there to working at typewriters or other women's machines. Eventually, studios' scenario staffs would be further subdivided into four or five departments, including writers, story, research, reading, and script/stenography. The final three were characterized by their largely female staffs, the aspects of their work that were clerical, and the atypical placement of women in many of their leadership positions.[158] Riddle also named nine female writers out of a total of seventeen in his 1921 description of the Lasky scenario staff, and certainly their continued success in screenwriting might account for their dominance of related departments, sometimes referred to as "literary" or "intellectual" fields. However, as later chapters explain, the inverse was true: women were able to continue their work in screenwriting, story, and the less impressive occupations of reading, research, and script or stenography, because (not in spite) of their association with typing and other aspects of clerical labor. This link explains women's continued participation as writers long after their prospects as directors, producers, and executives had evaporated. It also elucidates the women's typical domination and leadership of reading, research, and script departments, when women supervised story or writers' departments at the major studios far less frequently, and rarely, if ever, held managerial roles in any other creative, craft, or technical departments.

It is no coincidence that the three script-related departments women governed were ineligible for official acknowledgment for their contributions to the finished screenplay. According to rules that evolved around onscreen crediting for a film's written material, members of the remaining scripting departments—writers and story department personnel—might receive "story by," "screenplay by," or "written by."[159] These common threads—clerical duties and a lack of credit/creative power—connect script-related fields where women predominated with that of script supervision, which emerged as a women's job at studios around the same time. The same threads connect studio-era women's fields with the late- and post-studio-era reformulation of publicity, casting, and junior-level film and television development (itself a combination of reading, research, story, and secretarial work) as women's fields. The next chapter outlines the process by which female movie workers became associated with clerical and other preexisting categories of women's work as their workplace identity diverged more fully from that of the earlier film industry's powerful, creative female movie maker.

"HIT THE SWITCHES"

(Continued from page 3)

abstraction. If ever that gal concentrated for a moment she'd solidify. She was giggling and gazing over Bill's shoulder. "B. O.," she was whispering. "Isn't that vulgar, fancy a lamp having B. O."

"We're kind'a used to it. B. O., Minnie, means that the lamp is in bad order, it has to be sent to the shop for repair. Let's go over there, maybe I can fix you up with Harold Nye. He's in charge of the shop. Don't put your feet on that spider," we warned.

She shot into the air with a squawk. "Spider!" she shrilled. "Ooooo!"

We untangled. "What's in a name?" said somebody. He never met Minnie. We had to explain. "Spider is the name for that block of wood with those three bars running across it. It is used as a terminal for

"Who was that guy?" she demanded. "I'll . . ."

"Don't do any thing rash," we warned. "That was Ray Haslam. He's in charge of the 'iron' men."

"Iron men?" she queried, again with that look in her eye.

"Yes," we countered. "The men who go into a set after it is built and fix it up with lights and cable to feed them with. They're all tough guys."

"Oh!" was all we got. "What about this Mr. Nye?"

"In here," we answered, as we steered her into the 'shop'.

"This," we said with pride, "is the finest equipped shop in the business. They make old into new. Lamps, which in any other studio would be scrapped, are rebuilt and put into service. All new equipment is tested and given an O. K. or otherwise." We

a circuit. There's 220 volts across the two outer bars."

She looked a little impressed. "Let's get out of here," she wailed.

We passed some lamps on the sidewalk of Brownstone Street.

"Those are 5.k's.," we volunteered. "That means 5,000 watts each. There is about 1,600 of them on the Lot."

"And what's the little ones?" she said mildly.

"2.k's or deuces. Those with the whirly-gig reflectors are rifles."

"There must be an awful lot of lamps in the studio," she said, putting on some lipstick.

We did a bit of hasty figuring. "Yes," we returned. "Take the Inkies, that's the incandescent lamps. We have about 4,000 on the Lot. The arcs, 35 Suns, 18 of the big 36-inch and about 300 of the Mole Richardson type."

Just then the "hootie" swung round the corner, a blast from the horn and Minnie jumped into the shelter of the restaurant.

showed her some 65-amp spots. "These lamps were designed and built right here. They've already made about thirty. Then there's the 'headaches'. That's a converted lamp; it's the old 'deuces' fitted with a lens; they've got 1,600 to do and have 600 of them done. The big lamp there, they call it a 36-inch arc; you can read a letter in its beam a mile away. That bit of mechanism inside was designed right here by George Anderson. The Academy gave him an honorable mention."

Minnie just blinked. She was admiring the bubbles coming out of the plating tanks. "What are they boiling them things for?"

"They are plating them. They cover iron or steel with a film of cadmium or nickle. That's to keep it from rusting. It's the same thing as you have on the head lamps of your car."

She nodded her head as if she understood. "And is that a head lamp?" she asked, pointing to one of the new

miniature spots which Harold Nye designed.

"No," we returned, "that is for fitting on top of the camera Matte box. Harold Nye designed and built them."

"He must be clever," she smiled.

Honorable Mention

"Clever enough," we said. "The Academy thought so when they gave him an honorable mention in their awards bulletin."

"Aren't they pretty?" she added. "Look at the nice blue color."

It was Phil La Guess who "took the words out of our mouth". "Yeah. Mam," he joined. "That's Murphy blue. It was the late Frank Murphy who picked that color. You can see it all over the Warner Lot. I got gallons of it in the paint shop; come over an' see it."

Minnie looked at us and sniffed. "Who wants to see old paint?"

"We had better humor him," we cautioned. "These artists are touchy. Phil is in charge of the Electric Paint Shop, and covers a lot of territory with a paint brush."

Phil was at it again. "There's a sandblaster," he explained, pointing to a nozzle which was hissing like a thousand snakes. "We blast every bit of paint off the lamps before they are painted, and after the paint is on we bake it in that oven to a temperature of 250 degrees."

Minnie was shouting something above the hissing. We moved closer. "I don't like it in here," she was screaming. "It smells . . . and paint gives me hay fever."

There was nothing else for it, so we moved and passed by Bill Swank's dug-out. Bill wasn't in so we peered through the window at his litter of telephone gadgets. Minnie was blinking. She was about to question again. "That's the workshop for the repairs to the telephone system on the Lot. That also comes under the Electric department. We have our own exchange."

Number Please

"I used to be an exchange girl," said Minnie. "What kind of a board have they?"

"They have one large P. A. X. and a P. B. X. in the Publicity Department."

"I'll bet it's a nice job. Just fancy nothing to do all day but listen to the voices of the stars," she sighed.

"Nothing else to do?" we said. "No, nothing but handle the traffic of 700 stations with 50 outside lines and during a busy period handling calls from places as far apart as London, Paris, Melbourne or Calcutta. Marie Olsen is the girl responsible and she's got eight other girls working from seven in the morning until 12:30 next morning. We don't think you would enjoy it, Minnie."

We had been talking to ourselves for the last half minute and was only aware that we were alone when Minnie raced around the corner holding her skirts down.

"What's the matter?" we inquired with a grin.

"I don't know," she blazed. "Some-

(Continued on Page 15)

Photo by Buddy Longworth.
Telephone operators Doris Halligan, Opal Reel and Alice Ingersoll

ROBERTS TOLUCA LAKE DEPARTMENT STORE

5 - 10 - $1 Economy Department
and
Ladies', Men's, & Children's Wear
PLEASING SERVICE FOR ALL
10110 Riverside Drive
Telephone SUnset 2-0252

UNITED RADIO STORES

Since 1921　　Free Parking
The best place to buy your Zenith, P h i l c o, R C A-Victor, Emerson, Packard Bell RADIOS
Ranges, Refrigerators, Washing Machines, Electrical-Gas Appliances
Ask for . . . NORMAN
147 to 151 So. Bdwy. Phone MI-2233

Pete's Barber Shop

Tonsorial Art by Experts
For 14 Years at Old Sunset Studios
NOW AT YOUR SERVICE
—IN NEW STORE ACROSS FROM WARNER BROS. ADMINISTRATION BUILDING
3900 South Olive Avenue
Burbank, Calif.

● 6 - OLDSMOBILE - 8 ●

J. E. ROBERTSON
SUBURBAN MOTORS
822 East San Fernando Road
BURBANK
HE-8535 . CI-14262 . CH-62762

YOUR

HEALTH GUARDIAN
MAGNETIC
SPRING WATER

Phones:
BRadshaw 2-1183
OXford 1178

F——O——O——D

When You're
HUNGRY or THIRSTY
Day or Night,
MOTHER AND AL
have just the thing
IDEAL CAFE
4303 So. Olive

H—O—N—E—S—T—Y
C—O—U—R—T—E—S—Y

S——E——R——V——I——C——E

Fig. 10. Warner telephone operators pictured in the April 1940 *Warner Club News* feature "Hit the Switches." Courtesy of the Academy of Motion Picture Arts and Sciences.

2

Studio Tours

Feminized Labor in the Studio System

Before the cameras can start to grind an army of typewriters must do its work. These typewriters must be manipulated by skillful feminine fingers, which opens another field of endeavor for the girls who have hit the western trail.

—Ralph Wheeler, *Screenland*, May 1930

Once scientific management principles took over film studios in the 1910s, the practices of sex segregation and the feminization of certain types of film labor were sure to follow as female workers came to represent a particular type of labor to managers in industrial production systems, with a particular benefit to the bottom line. The previous idiosyncratic and individualistic practices under which, as Karen Mahar argues, women "enjoyed more latitude and leverage than women in any other industry, including the stage," dissolved as the movie industry began to view women's participation behind the camera as "archaic, and perhaps embarrassing, as the haphazard production methods of the nickelodeon era."[1] The hiring practices of other industries such as banking, where powerful women were not the norm, replaced the film industry's more progressive practices. This shift left only those sectors increasingly perceived as women's work open to female job candidates. Indeed, women's shrinking presence in leadership positions under what Mahar calls the "remasculinization" of filmmaking—a return to the all-male state of the first motion pictures, before women had a prominent place in the industry—and women's increasing association with the tedious, low-paying detail work around the margins of the filmmaking process were inextricably related.

Labor associations such as guilds, clubs, societies, and, later, unions reinforced many of these emerging divisions between male and female workers. Disputes over professional jurisdiction—which worker did which task—were often resolved by separating work into highly specific positions on the set, and then spreading the word about the job distinctions through trade journals and unions, which solidified their subdivisions.[2] Many early trade groups, such as the Screen or Reel Club, formed as fraternal organizations that socialized in masculinized spaces such as taverns and lounges. Other than the Writers Club, a social arm of the gender-integrated Screen Writers Guild with its own heterosocial clubhouse, most such groups eventually moved from the mixed gender spaces in which they had previously congregated to sex-segregated spaces like the Los Angeles Athletic Club. Barring the occasional ladies' night, women could not follow their male peers into these clubs and taverns to discuss how and by whom their jobs were to be carried out.[3]

Sex segregation at studios was, for the most part, de facto. No formal policy was necessary to limit women's roles in the workplace. The logic that governed their participation in other American industries—limiting them to work that was distinct from and less desirable than men's—reflected wider societal norms about innate feminine qualities and women's natural sphere. Growing studios enforced these norms more strictly than early film companies had done. As they increasingly associated women with certain work, studios effectively dissociated them from high-status creative fields in which they had found so much early success. The relatively heterosocial production environment of the early movie business, which allowed women some mobility between occupations, also became more restricted as job separations and hierarchy assumed a spatial dimension across studios' sprawling compounds. One could veritably map the territory of the studio grounds according to where women workers had access and the departments in which they were typically present or absent. By the 1930s, women's prospects in Hollywood would be narrowed considerably. All but eliminated from the ranks of directors, producers, and executives in which they had flourished a few decades earlier, women would be primarily associated with low-status jobs rooted in feminized labor sectors.

This chapter elucidates how the film industry spatialized and categorized women's labor as feminized sectors grew in size and number through the 1920s and 1930s. Extant labor history provides the traditional, accepted categories of women's work as they were understood in other industries and in American homes. Filmed studio tours produced for promotional purposes serve as a starting point for identifying these categories as they were assimilated by studios, where work was promptly separated along the same lines. Examining representations of the work and workers in these various branches of studio labor shows how women—as maids, commissary waitresses, nurses, teachers, inkers, painters, negative cutters, and film inspectors—were dislocated from creative work by being ranked lowest in studio hierarchy and farthest from the rolling camera in studio geography. However, the chapter also designates how important these workers were: studios could not have grown at the same rate or functioned on the same scale without women serving as a cheap, exploitable workforce to facilitate and subsidize their growth.

These examinations also necessarily broaden the chapter's scope to include labor assigned not only based on gender, but also on class, race, or a combination of the three, such as jobs that fell under that heading of "service" (custodians, maids, and so on) that were assigned to African American men and women. Little evidence relating principally to these workers was saved in archives, and what exists is difficult to locate because it is scattered unevenly over many collections organized around other topics, such as notable individuals, productions, or companies. This is perhaps even more the case with primary resources relating to nonwhite or immigrant studio workers than with the Caucasian women who are this book's principal focus; their work was valued and credited even less. Not only that, but when work was divided along racial and ethnic lines, the practice was rarely articulated in written documents and as a result, the ethnic or racial makeup of the workforce is sometimes only evident in photographs. The existing secondary materials that devote descriptions to these workers are mostly popular press books, which, to interest their broader audience, concern themselves with the minutia of life "behind the scenes" that scholarly histories of studio business practice and products often necessarily leave out (for example, who worked at the studio commissary, which stars had maids). These workers are nonetheless included this chapter's "tour" of studio labor to provide a sense of the array and scope of work that was carried out by women and other marginalized groups at studios and, more important, because their inclusion is necessary to answer the larger question this study asks about studio-era labor, which is *Who else?*

"A Lack of Avenues Leading to Direction": The Female Movie Maker on (and off) the New Studio Map

Examining the studio-era fates of the powerful, early women filmmakers from the previous chapter shows the change in management's attitude toward female employees during the studios' nascent stages in the 1910s compared to the 1930s, when it can be said that their large-scale production systems had matured into adulthood. Here, again, the work of Cari Beauchamp, Lizzie Francke, and others who have examined these early filmmakers' careers in great detail helps to explicate studios' developing conceptions of women's work and their impact on all women in the classical Hollywood era—from important individuals to groups of women in feminized fields.

By 1937, Dorothy Arzner was the only female director of the nearly forty directors employed at MGM. This was neither the first nor the last instance in which Arzner—a successful silent-era writer and director—stood as the lone female exception to directing's all-male rule. Ten years earlier, papers had announced her first directing assignment with a headline—"Lasky Names Woman Director"—in which her gender seemed at least as important as her job or the company for which she worked.[4] She was the first woman to join the newly formed Directors Guild in 1936, and remained its only female member until Ida Lupino joined in 1949. Arzner attributed her field's dominance by males not to men's superiority, but to the fact that women were "handicapped by a lack of avenues leading to direction."[5] And, indeed, directing—a difficult job to learn without on-set training and exposure to camera equipment and technology—was increasingly out of reach for women as not only film sets but also the technical and craft departments that emerged around them became the sole province of men. Most women working at MGM in 1936 were relegated to work with light machinery such as sewing machines and typewriters, and confined to all-female or female-dominated sectors shunted to the outer edges of studio geography.

Frances Marion's experiences as a contract writer for MGM in the 1930s and 1940s contrasted sharply with her earlier occupational mobility in the production systems of the 1910s, where she was encouraged to learn all kinds of film work. As an MGM screenwriter, she became frustrated as increasing specialization restricted her to the writer's building, with management discouraging and even disallowing her from further input into the production of films after she wrote them. She believed that she and other female writers had contributed to the movie factory that the studio system had become in diverse ways—none of which they received credit for.[6]

"Bess Meredyth, Anita Loos, and I were asked our advice on virtually every script M-G-M produced during the Thirties," she recalled in an interview with DeWitt Bodeen, adding that they concealed their power by carrying scripts in blank covers because they knew that "some male writers were complaining about the 'tyranny of the woman writer,'" and would have been embarrassed to discover Marion et al. were being asked to give suggestions and make uncredited revisions on their work. "It was a ridiculous accusation," she reflected in the same interview. "They were lucky to have us on their side."[7] Marion left MGM in 1937. Seeing her films marred by casting or production decisions on which screenwriters no longer had any real input had convinced her that the only way to maintain control over written work was as a hybrid: writer-director or writer-producer.[8] She successfully secured such a contract (to write and produce) at Columbia, but shelved the project after budget cuts across the studio's slate made it impossible for her to tell her story as she envisioned it.[9] No other directing or producing contracts followed, and an independent company she set up with Beth Meredyth and Meredyth's husband, Michael Curtiz, never got off the ground. Marion eventually returned to writing at MGM but served largely in an advisory and editorial capacity there, working with less experienced writers.[10] Meanwhile, Marion's old friend and former employer Lois Weber—once a brand name with her own studio—was, by 1932, consigned to work as a script doctor and at a "charity" job testing starlets for Universal.[11]

Marion's failure to either direct or produce for the majors—despite her status as one of the most successful screenwriters in history—gives some indication of the difficulties faced by women even with considerable standing in the industry at the time, let alone those with no standing at all. Though women often worked and even thrived in roles as screenwriters, and in the so-called literary or intellectual professions (for example, reading, research) related to the process, as Cari Beauchamp explains, screenwriting "was a creative outlet achieved in private and required relatively little bravado."[12] Women's success in more visible leadership roles was far rarer. Joan Harrison, one of the few women screenwriters promoted to producer in the 1940s, explained that it was difficult for a woman to succeed "except as an actress or, much down the scale, as a writer" because "the front office attitude resents a woman in authority and it probably always will— they recognize women writers but prefer to keep us in prescribed grooves."[13]

Though avenues to directing have reopened since Marion's and Arzner's time, still women directed only 6 percent of the top 250 films in 2013.[14] The field remains effectively segregated by gender, in part because guild membership rules require endorsements from three existing members, and current membership is still dominated by white men. More broadly speaking, a 1998 study of female-dominated professions pointed out that women are often segregated into female-dominated sectors where they work in isolation (for example, the teacher in her class, the receptionist at the front desk) and are undervalued "either in terms of salary, career prospects or social status."[15] At studios, assigning women to literary or intellectual fields was often framed as a compliment in trades and promotional materials from the 1910s and early 1920s—a backhanded one, to be sure. Thus the literary fields—with their timeworn associations between women and typewriters—afforded opportunities for women to succeed, at the cost of effectively distancing them from the site of real creative power: the masculinized spaces behind the camera and in executive suites. Bluntly put, in a world in which the medium of *film* reigned, the agency of female movie makers was now confined to *paper*, along with all of its pejorative association with clerical work. This arrangement codified the logic according to which women's labor was below the line and at the margins of studio lots, allowing studios to exploit their workers by recourse to a self-explaining system of traditional gender norms.

Behind the Screen: Sex Segregation on the (Real and Imagined) Studio Lot

As chapter 1 detailed, early studios engaged in near-continuous self-promotion as they built their production facilities, "plants," or "works," during the mid-1910s and 1920s through the release of photographs, drawings, and maps in trade and fan publications. These materials were often packaged with accounts of reporters' (clearly staged) visits to, or tours of, studios-in-progress. Many studios also produced short films about their own facilities, works that likewise highlighted their ownership and mastery over the entire production process, reassuring potential viewers—from fans to competitors to potential investors—that they were concerns worthy of investment capital and consumer loyalty. Much like their print counterparts, these pseudo-documentary shorts, released mostly between 1915 and 1925, were framed as tours. Given their promotional agendas, these shorts are by no means records of what was. However, viewed as self-portraits, the films are revealing in both what they display and what they withhold from view, providing rich evidence of studios' gendered labor logic.

All of the tour films were planned, choreographed, and staged for the camera to varying degrees. *Behind the Screen*, made in 1915 and existing only in partial form today, gives a fairly candid backstage view of Universal City as a producer preps a film for production, while a later film entitled *Universal Studio and Stars* (1925) provides a more staged glimpse of the studio, particularly of its upcoming releases in production.[16] *A Tour of the Thomas H. Ince Studio, 1920–22* (released in 1924) and 1922's *A Trip to Paramountown* contain a mixture of candid behind-the-scenes shots of workers in various departments and sequences that are far more "produced," presenting carefully composed, even scripted, bits of business on the studios' sets and back lots.[17] Some films "map" the production process, starting with planning or scripting departments and ending with editing, thus reproducing studio workflow and chain of command. Others, framed as walking tours, reproduce actual studio geography. MGM's 1925 *Studio Tour* does both, and is perhaps closest to a straightforward industrial film about a factory and its contents. In early shots, the camera moves through MGM's front gates and then navigates around the lot, offering seemingly candid shots of facilities and the workers within them. However, the MGM tour also includes a "class picture" shot of workers from each department, assembled on the lawn in front of studio administration buildings.[18] A 1927 Goodwill Pictures short entitled *Life in Hollywood* features scenes from a number of studios, including footage of major Warner Bros. and Fox stars and directors.[19] Produced in the classical studio era proper—much later than the other films—the 1934 Warner Bros. short *A Trip Thru a Hollywood Studio* and a 1935 tour of 20th Century Fox present more mature self-portraits, in which feminized labor is fully incorporated into the larger studio system.[20]

Unlike the print articles discussed in chapter 1, which mostly implied feminized labor sectors through maps and photographic arrays of various buildings and assets, the filmed tours all include photographic representations of actual film workers, including some female workers deemed beneath mention anywhere else. Conversely, in presenting studios to (what the filmmakers believed to be) their best advantage, the films frequently omit important female creative personnel, most notably women directors. The appendix to this book lists the representations of labor in all films, categorizing the workers who appear by both gender and the department or job in which they are shown. This compilation of data is not intended as an accurate measurement of how many women and men worked in these departments. Rather, the ways in which the docu-advertisements included women indicate the fields with which they were being identified at that time. Following is a list of all jobs in which women appear:

Actor	Film Inspector
Art Director's Assistant	Film Lab Worker
Copyist/Typist	Hairdresser
Costume Designer/Department Head	Maid/Servant
Cutter	Mail
Dance Instructor	"Mannequin"/Model
Dancer	Modiste Assistant
	Modiste (Dressmaker)

Negative Cutter	Scenario Department
Note-taker	Scenario Writer
Nurse	Screenwriter
Patcher/Assembler	Script Clerk
Pianist	Seamstress
Publicity	Wardrobe Assistant
Reader	Wardrobe Mistress
Researcher	Wardrobe Staff

"The Antics of the Birdies": Women Movie Makers as Represented by the Tour Films

All of the tours display *some* women in high-status creative roles. However, the majority of women in this category are stars—the studios' most prized, oft-displayed human assets and, by the studio era, the personnel with the least agency of all major players in the creative process. Though extras and most dancers or models were hardly powerful players, for purposes of clarity this discussion categorizes them with stars as "performers" because their front-of-camera roles set them apart from other lower-status women's jobs in how they are represented.

Anthony Slide has said that women "virtually controlled" the silent film industry. He attributes much of that control to the influence of female stars with production companies, saying, "Certainly such companies might be managed by men, but if, say, Gloria Swanson chose Joseph P. Kennedy to manage her company, that in no way detracts from Ms. Swanson's power."[21] Perhaps not, but neither does Swanson's appointment of Kennedy as manager of her company indicate that legitimate power or authority was being vested in Swanson and women under this developing system. Genuine professional power exists and circulates in the open and carries unequivocal, discernable career capital (for example, additional job titles, salary, rank, position in the chain of command) with an implicit trade value in the industry at large. And if women as a group really possessed such power in the late silent industry, they would have had widespread placement in major leadership positions as studio managers, executives, or department heads. But this was hardly the case.

Though most, if not all, of the studios in question employed women writers throughout the 1910s and 1920s, female scribes appear or are referred to in only three of the films. The partial Universal film from 1915 includes a visit to "The Scenario Department," in which four men and one woman are pictured sitting on one side of a long conference table. The MGM tour shows three women (Agnes Christine Johnston, Jane Murfin, and Fanny Hatt) among the group of seven MGM screenwriters presented, so as to illustrate the claim that "some of the foremost writers of the day contribute original stories to M-G-M." Roughly eight out of eighteen total scenario writers are female.[22] The Fox tour ten years later includes shots of four female writers (Bess Meredyth, Sonya Levien, Helen Morgan, and Gladys Lehman) out of a total of twenty screenwriters, department heads, or story editors shown. Women writers are absent from the other films, which perhaps owes more to the status of writing at the time than to a refusal to show female writers onscreen.[23] Writing is omitted altogether from the Paramount and Ince films. Instead, on the Ince tour, the planning process is represented by a shot of Ince himself "as he confers with his production staff," which consists of seven men in suits working at a conference table behind Ince's own large, wooden desk. This masculine tableau seems a fitting representation of executive-managerial power at efficient studios like Ince's.

Scenario was one of the few departments where leadership positions were relatively gender integrated in the late 1910s. In their overview article for the Women Film Pioneers Project, Jane Gaines and Radha Vatsal list twenty women known to have worked as silent-era scenario editors, several of whom worked at the studios that would become majors and minors after the coming of sound in the late 1920s.[24] June Mathis led Metro's scenario department in 1919 before accepting a similar position as editorial director at Goldwyn, and later, Famous Players–Lasky. And at Universal, Eugenie Magnus Ingleton ran the scenario department in the late teens with five female staff writers out of fourteen. However, women's solid footing was hardly assured moving forward into the era of specialization and the conversion to sound. Mark Cooper points out that Ingleton was the only female production department head at Universal in the late teens, despite the popular notion that women ascended to high positions there based on job performance rather than gender.[25] By the 1920s women's occupancy of scenario department leadership roles began to wane, though many continued to write important screenplays and remained key assets for studios. Unfortunately, their talent or body of work—though important considerations when assessing the writers' legacies years later—did not necessarily translate to the kind of direct power that even Frances Marion found herself in search of by the 1930s.

Indeed, despite the roles women had been playing in high-status production fields in the 1910s, no female directors, producers, or executives can be seen on any of the tours, and none is identified or referred to, either. Male moguls, studio heads, producers, and executives frequently appear conferring over studio assets, and male directors are shown managing male crews in production. The absence of female management personnel accurately reflects the lack of female executives at many studios. As for female directors or producers, if any were employed by the studios in question (as they still were at some studios into the early 1920s), they were not selected for front-and-center display in the tours. Thus, in these representations of creative and managerial power, women are limited to spaces in front of the camera as talent or behind typewriters and in the minority at writing departments. Women's lack of representation in movie maker roles on the tours illustrates studios' shift toward staffing and representing themselves as big businesses with "serious" practices, thus by necessity managed and controlled by men.

Danae Clark describes typecasting in the studio system as a practice in which "actors were categorized according to social types, based on race, age, sexual stereotype, and so on," the result of which was fragmenting actors' labor power "by limiting their range of performance and preventing the full potential of their skills."[26] Allowing gender-based decisions to eliminate half of the directing candidates narrowed the range of creative outcomes, especially as time went on and the field masculinized further. Typical directors were not just white, heterosexual, middle-to-upper class, and male but also, styled after Erich von Stroheim and Cecil B. DeMille, decidedly masculine and authoritative with commanding manners and interests in hunting and other rugged, outdoor sports. Though DeMille consciously crafted his masculine personae, this archetype came to dominate both public and industry conceptions of what a director should be.[27] Perceptions also developed about which kinds of people were suited to the role of producer, writer, designer, and so forth. Dalton Trumbo explained the benefit of such typage to management, saying, "The front office likes to type writers, actors, everybody. It saves them having to do any original thinking, which isn't their strong suit."[28]

Typecasting of creative leadership roles likely made film production more predictable and manageable while simultaneously reducing its artistic potential. Many other areas of production labor were typecast in similar fashion and the practice has continued to this day, only interrupted when management has perceived some benefit in hiring new and different types for the role. In the case of directors, such a shift came in the late 1960s, when a younger generation, perceived as outside of the mainstream, was hired to break studios and their aging cadre of house directors out of a box office rut. This ultimately led not to a more heterogeneous workforce, but a new director "type"—that of the young, passionate artist, a product of film schools and counterculture and, yes, white, middle class, heterosexual, and male. The homogeneity of certain roles remains a fact within production culture and its mythology today, a testament, perhaps, to the course that efficiency set not just for women, but to some degree for all film workers.[29]

Many of the filmed tours show relatively heterosocial creative sectors (the scenario department, on-camera talent), but still tend to emphasize femininity over the individual talents or skills of female workers in these higher status roles. Nearly all of the films present some sort of feminine "eye candy" to sex up explanations of the production process. Shots of scantily clad female performers are typically delivered between intertitles that all but catcall the women onscreen. The Ince tour includes scenes of a fashion show in which studio-made garments are displayed on "mani-kins" or live models on a runway. *A Trip to Paramountown* visits the production of *Her Gilded Cage*, in which a number of dancers in short, glittering costumes perform on a stage before director Sam Wood and his camera crew. This segment presents several cutaways to the film's tux-edoed, male cast members, who are not acting in the scene, but standing to one side of the camera, smiling, watching, smoking pipes, and—per an accompanying title—enjoying "the antics of the birdies." The MGM tour contains similar bits of business, including shots of dancers kick-ing in a chorus line and a dolly past showgirls getting dressed in a scene from a "backstage" musical. Ten years later, the Fox tour shows a simi-lar lineup of female dancers in shorts and tank tops, trying out for male casting and dance directors, prompting narrator Jimmy Fidler to inter-ject, "Ahem, those you don't want, boys, I'll take," as the men onscreen turn conspiratorially to the camera. None of these images is surprising in the ritual of asset display. It is simply one link in a chain of evidence that female workers at studios, regardless of their status, were displayed (and thus regarded in studio culture) primarily in terms of gender and sexuality. Although they helped the studios sell products, these scenarios came at very real personal cost to the "birdies" involved, as evinced by the studio-abetted sexual assaults of dancer Patricia Douglas and singer Eloise Spann by male MGM employees in the 1930s, and the atmosphere of sexual harassment described by other former chorus girls, which is discussed further in chapter 3.[30]

Following the same gender-normative line of reasoning as the chorus girl scenes, *Life in Hollywood* presents actress Bessie Love sitting next to a motion picture camera, sewing between takes, explained with the inter-title: "If needle-work interferes with your art, stick to your needle-work, is Bessie Love's axiom." Love demonstrates her commitment to this motto when male crewmembers ask her to resume shooting and she resists, seeming to prefer sewing to acting. Though likely staged to amuse, this bit of business is a fairly accurate reflection of what, by the 1920s, was becom-ing a typical notion of women's place behind the scenes in the developing studio system. For, if these tours implied women were unacceptable or not the right type for some roles, they classified women as the *only* acceptable candidates for others. As such, in keeping with practices from other industries, female workers in the tours were frequently shown as segregated from men in departments that specialized in gendered labor not far off from Love's needlework.

"Stick to Your Needle-Work": Studio Adaptations of Preexisting Women's Fields

Areas in which women's work was deemed acceptable in major industries derived from essentialist views of women's natural sphere and innate qualities and skills. Studio adaptations of these basic women's labor sectors shared common characteristics, not only ensuring that these jobs would retain their gendered status after transplantation to film production, but also affecting which new, film-specific labor would later be typed as women's work, even when the connections to traditional notions of femininity were less literal. The first of these common characteristics was the work's lack of appeal to male candidates, based on cultural norms, compensation, and status.

In *Out to Work: A History of Wage-Earning Women in the United States*, Alice Kessler-Harris describes a hierarchy of desirability within the emergent class of nineteenth-century women's fields. The hierarchy reflected a cultural emphasis on the goals of marriage, home, and family by defining vocational success for women "in terms of values appropriate to future home life: gentility, neatness, morality, cleanliness," and "affirmation of home roles."[31] These values were diametrically opposed to the values encouraged in young men in the workplace, such as "ambition, competition, aggression," and acceptance of risk in the interest of upward mobility. Under such gender-normative workplace values, many of the jobs deemed least appealing by and for men were deemed most suitable for women. By aligning sociocultural with industrial-economic imperatives, the implicit values system rendered women's presence in the workplace less threatening to men's prospects there. In an alignment between sociocultural ideologies and economic realities, these values guided women to the very jobs that, in an era of efficient mass production, cost too much (because of the sizeable labor force required) to staff with white male workers at higher wages. Bound up with notions of certain jobs' cleanliness and gentility or lack thereof were notions of race, class, and ethnicity. Both women and men from ethnic and racial minorities were hired into those jobs least desirable to white, middle-class, and working-class women.[32] The other stops on this chapter's tour of women's work will elaborate and explain these categories of acceptable women's labor as they were adopted at studios.

Domestic Service

Under cultural and social norms, domestic service roles, particularly those related to cleaning, were perhaps the least desirable of the jobs open to women (and largely refused by white men because of their domestic ties and service aspects). Yet they were also some of the most numerous in America during the 1910s and 1920s. According to one manual citing Department of Commerce and Labor statistics, out of a total of 4,833,630 women working in 1909, 1,953,467 held positions in the fields of "domestic and personal service."[33]

By the 1930s, the major studios had grown to the size of small cities, complete with cities' infrastructure, social strata, and need for service workers. As Beth Day explains in *This Was Hollywood*, her early, popular history of the studios:

> Here the thousands of workers who poured into the carefully guarded gates each day found the necessities of a highly urbanized life. There were the private electric and water systems; the private telephone exchange; the barbershop and drug counter and magazine stand; the bootblack and the dentists, the doctor, the nurse and infirmary. . . . There was a giant commissary equipped to feed thousands which also provided special arterial service to the private dining rooms ruled by the company's top executives and key producers. Also, for the elite of these caste-conscious worlds, there were such added refinements as a heated swimming pool, a Turkish bath staffed with a fleet of masseurs, a sundeck, and a vitaminologist who did nothing but administer shots to high-salaried posteriors.[34]

Studios employed not only white women, but also immigrants and people of color of both genders in service roles. These gendered and/or raced workers sometimes adorned published photographic spreads that, like the filmed tours, promoted studios as cities, complete with nonwhite and female workers in their traditional serving roles. Identifying such workers' racial or ethnic backgrounds in grainy, aging films and photographs, and in statements from studio workers containing retrograde terms regarding race, is a fraught and uncomfortable process; and yet, it is more fraught to simply omit race and ethnicity from this discussion of low-wage, low-status sectors.

Several tours show women and minority men working as butlers and maids. For example, at the Ince studio, a woman in a maid's uniform cleans the wardrobe department, while at Paramount, a similarly uniformed woman attends Gloria Swanson in her dressing room. A butler who looks to be of Asian ethnic origin waits on producer Thomas Ince in his home before the day's work in the Ince studio tour, while at the studio proper, a uniformed black or African American male employee distributes packages and mail to arriving stars.

African American and Caucasian women worked as maids at a number of studios in the twenties and thirties. At Mary Pickford's United Artists Bungalow, "a butler and a cook were always on duty," and Marion Davies's bungalow at MGM was "run by a full staff of servants" catering not only to Davies, but also to studio executives who entertained distinguished guests there. And in interviews and oral histories, a number of major creative personnel recalled individual maids. For example, costume designer Walter Plunkett described a young "black girl" serving as maid to an actress or in the costume department at silent-era studio Film Booking Offices of America (FBO) in 1926.[35] Rosalind Russell recalled that at MGM, a maid named Hazel Washington attended Greta Garbo before Russell later hired her as her own maid.[36] Some stars hired domestic service workers themselves, though their wages might be paid by the studio. An African American woman in a maid's uniform is also pictured with Jean Harlow in *M-G-M: Hollywood's Greatest Backlot*, with the caption "Jean Harlow and an assistant walk past a sounds stage on Lot One 1930s."[37] The photograph and caption are a good example of why, absent photographic evidence, it is often unclear who did these jobs, and that race played a role in who was assigned what work. These workers may not have participated directly in the making of films, but they were nonetheless a part of studio life at the time.

Growing studios also employed janitors to maintain order and cleanliness. Though not represented in promotional films, janitors are described in the personnel column of the *MGM Studio Club News*, that studio's employee-authored house organ, and male custodians and their female counterparts, matron custodians, had their own column in the equivalent newsletter at Warner Bros. and Fox, indicating that at those studios, such jobs were performed by full-time studio employees, rather than by independent contractors.[38] Horace Hampton wrote the janitors' column, "Sweepings," for the Warner newsletter in the early 1940s. In January 1940, he joked that Warner employees "come to work each morning with a faith that is phenomenal" (referring to the workers' expectation that the messes they've left on their desks and chairs the night before would be cleaned up), adding, "The Janitors' department is eager to supply these services, however, the degree of benefit derived not only depends on the Janitors in fulfilling their duties but also to the manner in which employees accept their services," an oblique reference that Hampton follows up with a final, direct plea for employees to help keep the workplace clean.[39] A year later, the newsletter carried a cover story on the workings of the custodians' department, which included photographs, and all of the several dozen employees pictured appear to be African American or from another racial or ethnic minority. The department, the story explains, had grown from a staff of two in the early days of the studio to fifty-five people, roughly a fifth of whom were women, based on photographs and descriptions of their shifts (which include twelve men and three women).[40] At Fox in the 1940s, the

Fig. 11. April 1941 *Warner Club News* cover story on the custodial department. (Courtesy of the Academy of Motion Picture Arts and Sciences.)

service department included roughly fifty-seven janitors working on both day and night crews, along with five "matrons" and three window washers.[41] In the Fox house organ, *Close-Ups*, custodian Ted Crumly calculated, "There are over three hundred desks in the Administration Building Alone and approximately one thousand windows. Almost one-half ton of waste paper is removed from this one building daily."[42] Based on this evidence, it is unclear whether the laundering of clothes and other fabrics was contracted to outside cleaning services. Personnel documents and employee accounts that describe studio infrastructure do not include laundries or laundry workers in their maintenance, custodial, or service departments.

Studio barbershops were another service sector employing African Americans. At MGM, the barbershop and newsstand was a gathering spot, and a 1932 *Fortune* magazine profile of the studio described the shoeshine parlor, which was staffed by an African American man or, as the author put it, "the colored shoeshine boy" who frequently earned "a day's pay in an African mob scene," and also "worked as chauffeur of one of MGM's sixteen company limousines."[43] This passage may have referred to Harold Garrison, who "was known for his skill at accompanying his work with song and dance routines."[44] Garrison chauffeured Irving Thalberg and other studio executives, and was a reputed confidante of executives and stars.[45] A similar shoeshine parlor/barbershop arrangement could be found at other studios.[46] Howard William Washington ran the shoeshine stand at Warner Bros. in 1940, and oversaw another worker, A. C. Johnson, an African American man pictured in the October newsletter, giving "Officer Bob O'Neal" a shine.[47]

By all accounts, studio barbers were white and male—like the two featured on the MGM tour. Even studio makeup departments were masculinized in line with the rest of the beauty industry, and leadership of such departments was dominated by men, particularly George Westmore and his six sons, who between them at one point led departments at all of the major studios but MGM, where Jack Dawn supervised makeup.[48] In *It Takes More Than Talent*, Mervyn LeRoy's 1953 how-to book on breaking into the movies, the MGM director addresses the question of whether the field is open to women by saying that "ninety per cent of make-up artists are men."[49] These figures are borne out by available descriptions of studio makeup departments, for example, in the *Warner Club News* cover story on the makeup department, which lists full-time employees and their jobs. All of the men mentioned in the article are makeup artists, while all the women are hairstylists. The field of women's hairdressing, typically classified as a subdivision of makeup departments, was female dominated.[50] Like the barbershop, the women's hairdressing trailer served as a social gathering spot for female stars at studios like Paramount, which had one large

Fig. 12. October 1941 *Warner Club News* cover story on the annual "informal" dance held by the custodians and service workers. (Courtesy of the Academy of Motion Picture Arts and Sciences.)

space for all hairdressing. According to Ronald Davis, hairdressing was the first stop of the day for most actresses at MGM, and the place where "gossip was bandied about" concerning the previous evening, ongoing productions, even domestic squabbles.[51] Nellie Manley, who started the hairdresser's guild with thirteen other women, said actresses often preferred to have the same hairstylist every day because "you become rather personally involved with them," and as the first person they would see in the morning, "you become quite the confidante. It's almost like being a wet nurse!"[52] In her oral history, actress Fay Wray recalled that women who were close to the set "were very wonderful and supportive and very attentive" and echoed the notion of hairdressers as confidantes that actors didn't mind telling about their personal lives and problems because "they will protect you," adding that the woman who headed the hairdressing department at Columbia was known as Mrs. Columbia to actors who "depended on her kindness and attention and concern about anything they were doing."[53]

Front-of-house restaurant service was still considered men's work in many fine-dining establishments around the turn of the twentieth century—a vestige of practices in large, upper-class households of the past—as well as in masculine spaces such as saloons. The commissary waiters shown in the 1915 Universal tour, *Behind the Screen*, are white, male, and dressed in black coats with white bib shirts. Likewise, a male waiter is visible in a picture of the Ince commissary in 1918.[54] As time went on at studios, white male commissary servers typically remained only in positions of higher status, such as headwaiter or majordomo. Nick Janos was headwaiter at the Fox commissary, having come there from a similar position at the Brown Derby at the request of his friend Darryl Zanuck.[55] Studio boss Jack Warner and his executives ate meals in a private dining room with its own white-gloved butler.[56] A similar arrangement existed at MGM, where Louis B. Mayer and his executives ate in a private room that included a Directors' Table where, for seven years, waiter Billy Fies served the same, all-male group of high-profile actors, directors, and executives. They played dice to determine who would pay his tip.[57]

However, food service jobs were also open to women at studios by the 1910s, and female servers were the norm by the 1920s, as reflected by the white-aproned waitresses among the studio's commissary staff on the MGM tour. Women were the only uniformed servers captured in photographs of the Warner Bros. commissary in 1937 and the 1940s.[58] A Paramount employee described twenty-five waitresses working there in 1936.[59] Workflows from the 1940s record no waiters, but roughly forty "waitresses," a "telephone girl," and two "counter girls" at work in the Fox restaurants.[60] Female waitresses were the norm at the MGM commissary as well; they formed the first female bowling team in the studio league in 1938, calling themselves "The Slingers," beating a team of male competitors in their first match-up.[61] However, though these waitresses were on the studio payroll, according to the *MGM Studio Club News*, the restaurant that opened on Lot 2 in 1938 was run by the Brittingham Company, an independent firm that also operated pushcarts selling candy and soda around the lot, staffing both with its own, outside employees.[62]

Of course, men of color and white adolescents and young men carried heavy trays and cleared tables as commissary busboys and are visible in MGM's *1925 Studio Tour*, the Ince tour, and in archival photographs of the studios.[63] These jobs fell inside the narrow range available to men of color on studio lots, where black workers were subjected to the same

Fig. 13. December 1941 *Warner Club News* cover story on the studio commissary workers. (Courtesy of the Academy of Motion Picture Arts and Sciences.)

systemic and cultural racism pervasive throughout the country at the time. Though the Jim Crow laws of the South did not extend to Southern California studios, most American businesses nonetheless observed a tacit policy of workplace segregation, under which it was understood that black men would not be hired in most positions and would be the *only* type of employees hired for others. Segregation extended to commissary clientele in at least one studio. When Lena Horne and other black cast members from *Cabin in the Sky* were refused seating by L. B. Mayer's brother Jerry, the studio manager at the time, L. B. opened his private dining room to them.[64] Until that time, at that studio at least, black movie workers might have worked in the commissary, but they were not served there.

At studios, the commissary was one of the few places where women typically supervised others' labor. The supervision of restaurant service labor in America was, by the early 1900s, one of the few managerial roles deemed "feminine" enough to be suitable for women, and undesirable enough—because of the service aspects involved in managing personnel while catering to clientele—to be unattractive to white men.[65] Though the aforementioned Nick Janos managed Fox's Café de Paris commissary, the dining room's "head hostess" oversaw other dining room staff in Fox workflows.[66] Katherine Higgins managed the Warner commissary in the 1930s and was responsible for catering meals to stages in production. She conferred daily with assistant directors in order to supply food on set, and her department provided 120,000 boxed lunches yearly for location production.[67] Pauline Kessinger performed similar duties as manager of the Paramount commissary, where she remained for thirty-five years, rising from a waitress position in the 1920s. She also supervised the backlot café and coffee shop, managing a staff of eighty waitresses, busboys, cooks, and dishwashers, some of whom worked late into the night to make as many as fifteen hundred lunches for location crews who might pick them up as early as 4 A.M. Kessinger's other duties included attending production meetings every day to know how much coffee to cater to sets, convening regular instructional sessions on service, and swapping in "new uniforms for the girls" every six months when customers "got tired of seeing the same old dresses."[68] Meanwhile, at MGM Frances Edward was, per that studio's newsletter, "boss of the commissary" by the late 1930s, and in the summer of 1940 the newsletter reported that the "commissary cuties" would "blossom out in brown and white" uniforms that were being changed to match the color scheme in the dining room.[69]

Commissary spaces were generally not feminized beyond their service jobs. Positions in the kitchen (other than those involving baking) were masculinized in America in the early twentieth century, as reflected by male chefs, cooks, and dishwashers at the bottom end of the kitchen hierarchies, the "pantry men" and other gender-specific jobs on the Fox flowcharts, and the mentions of only male chefs in newsletters.[70] And though Katherine Higgins managed Warner commissary personnel in the 1930s, the kitchen staff worked under the direction of Charlie Bader, who headed the food department to Higgins's service department.[71] Still, commissary and food service work was a decidedly feminine operation, with Kessinger and others cast as "den mothers" charged with not only serving good food, but also providing an inviting atmosphere and making outsiders and regulars alike "feel they were the most important of all" in studios full of VIPs.[72] At MGM, L. B. Mayer had his commissary serve his mother's chicken soup to create a family atmosphere and, it was hoped, dissuade workers from the inefficient practice of leaving the lot for meals.[73]

Domestic Arts and Crafts

Women's labor was culturally acceptable in areas related to domestic arts and crafts, including the middle-class women's parlor art of piano playing, and so-called home arts such as sewing and needlepoint, often used for clothing or interior decoration. While men can be seen playing violins and other instruments in some of the studio tours, apart from one female violinist pictured with members of the orchestra units at MGM, the only two female musicians among roughly thirty to thirty-five total musicians pictured in all the films are seated at pianos.

However, though their representation was scarce, inconsistent, or even nonexistent in so many other sectors of labor, in every studio tour in which the costuming process is shown women can be seen engaged in work related to the design, construction, and fitting of clothing as costume designers, wardrobe mistresses, and assistants. In the Ince film, only women fit and manage the costuming process and only women sit at sewing machines. However, the presence of male wardrobe assistants and designers in the Universal, MGM, and Fox tour films attests that, by the 1920s, only certain phases of costume production were considered strictly women's work: those with domestic rather than artistic associations (for example, direct work with a needle and thread) that were of low status in the creative process and hierarchy, and that involved repetition or what might be called detail work such as embroidery, making use of women's "natural" dexterity, tidiness, and aptitude for carrying out routine tasks.[74]

Feminization did not extend to aspects of costuming that were considered to be of creative or managerial importance—namely, designing or supervising. Women had occupied these roles in silent-era costuming, and in her 1922 essay on the field, Christie Film Company designer Edith Clark characterizes clothing design as a women's field. As studios grew in the 1920s and '30s, costuming remained more open to women than other areas of production design at studios, such as art direction or set design.[75] And certainly, many of the most successful and well-known costume designers of the classical era were women. The notability of these female figures and the female-dominated state of costume design today has fed the perception that this sector was a women's profession in the studio era as well. However, costume designers and heads of departments were at least as likely to be men as women, if not more so.

At Paramount, the studio's head costume designers, Travis Banton and Howard Greer, outranked Edith Head until the late 1930s, when—according to Linda Seger—she became "the only woman to oversee the design department of a major studio."[76] The mononymous Adrian was costume supervisor at MGM until 1941, when Irene Lentz Gibbons took his place, joined later by Walter Plunkett and Helen Rose. Orry-Kelly (the professional name of Orry George Kelly) was the major designer at Warners for much of the 1930s and 1940s, and Edward Stevenson served as head designer at RKO Studios from 1936 to 1951. At Fox in the 1930s, Gwen Wakeling shared supervision of wardrobe (which at Fox included costuming), with Arthur Levy and Royer (also mononymous), while Charles LeMaire headed the department in the 1940s.[77]

Many female designers worked in these departments, but men were more likely to hold top-level positions at most studios, which were at best evenly split by gender until the late 1940s. MGM's *1925 Studio Tour* makes clear that although wardrobe mistress E. F. Chaffin and her staff execute costume designs, the designs themselves originate "with the great Romain de Tirtoff, Erte, the world's foremost designer," who is depicted in his own scene, fitting one of the studio's stars amid lush drapery. Other male workers are among the wardrobe department's staff when arrayed for their class pictures on the MGM lawn, but they do not appear in subsequent scenes of the workroom, where only women sew with needles or at machines. Male wardrobe workers appear in subsequent shots of the men's wardrobe department, where they act as clerks, checking men's wardrobe in and out to male cast members. No sewing machines or other tools are evident in the men's department, and no workers are sitting down, working on costume production. It is unclear whether men in MGM's wardrobe department typically performed any costuming labor, but, significantly, their work as depicted in the films is free from those duties associated with home arts.

A similar state of affairs is reflected in the other tours' displays of male wardrobe and costume workers, which either show them in postures that reflect supervisory duties (standing, conferring with managers while women sit working at machines) or specifically identify them as costume designers, fashion designers, or departmental supervisors. At the same time, of the roughly fifty women to be seen in the same films' costume and wardrobe departments, only three—the MGM wardrobe mistress, the Ince wardrobe mistress, and Gwen Wakeling, co-supervisor of wardrobe at Fox—are identified with supervisory or creative design work. The other forty-plus female workers are either assisting in fittings or doing the sewing labor of hand and machine stitching. In the world of film production as these films present it, the design of costumes (where all artistic and production value resides) is a masculinized or mixed-gender profession, while the actual labor required to produce costumes is women's work.

This filmic self-representation of costuming by studios is accurate in its division of labor at studios' wardrobe departments, which expanded throughout the 1910s and 1920s along with film production to include "a chief designer who was assisted by the head of wardrobe, several junior designers, sketch artists, period researchers, wardrobe assistants, and seamstresses."[78] The work in these departments included sewing, dyeing, beading, millinery, and many other subspecializations, which, together, allowed studios to costume any role from head to toe.[79] Describing the Lasky fashion department in the 1920s, studio manager Melvin Riddle states that 100 to 125 "girls and women" were "employed at all times" in Lasky's fashion department in the early 1920s, including thirty or forty seamstresses at sewing machines.[80] Men were less commonly employed for lower-level costuming work outside of a few specific roles, one of which was wardrobe assistant or clerk, like the male wardrobe workers displayed on the MGM tour. In earlier years, studios' male stars brought their own wardrobes to wear in films with modern settings, and male wardrobe workers were present on set largely to assist male actors rather than to create their garments.[81] The practice of men dressing men continued as costume designers began to design male stars' attire along with that of their female costars. There were also a few masculinized specialties related to costume production, such as menswear tailors and drapers (who also worked in set design). Male tailors can be seen constructing suits in photos from MGM's costume department in the 1930s.[82] On the whole, though, staffs of workers responsible for constructing and maintaining collections of costumes were largely made up of women throughout the 1930s and 1940s.

Costume production's low status and low level of creative agency, as well as the domestic associations of the process, contributed to perceptions that the work was for women. Fueling this perception was the work's association with repetition, detail, and tedium, as required in sewing with a needle and thread, cutting out patterns, and so forth—traits that, combined with the domestic associations of wardrobe's requisite cleaning and maintenance of clothes, meant such work was under the purview of women.

Indeed, some parts of the costume production process were so specialized, repetitive, detailed, and—in some cases—mechanized that they might be more accurately be grouped under the next category of traditionally feminized labor on this alternate studio tour of women's work.

Light Manufacturing

This feminized sector of American industry arose in the 1800s when factories began mass-producing wares that women had previously supplied to their own families through home manufacture and "taken in" for wages (for example, textiles, piecework). Home manufacturing associations sanctioned the hiring of women for corresponding specializations within factory production. The work was deemed light enough for women (and, frequently, adolescents) who, conveniently, were also considered better suited to such up-close, detailed work than were higher-priced male workers.[83]

Even after women's manufacturing jobs became mechanized and the labor ceased to resemble that of home manufacture, the link among women, fine-detailed work, and light manufacturing persisted. And so, women were associated with sewing machines in factories just as they were associated with typing machines in offices. It is this kind of light manufacturing in which female wardrobe and costume workers are most frequently engaged when they appear onscreen in the studio tour films. Furthermore, these workers are depicted in strikingly similar ways in all of the studio films that show actual costume production (Universal's *Behind the Screen*, and the Ince, MGM, and Fox films): three or more women in near-identical postures, bent over sewing machines, seemingly unaware of the camera, their fellow workers, or anything other than their work. Depicted in this way, the workers are mere extensions of the technology they operate: mechanical assets rather than human resources. Similar tableaux were standard in trade and fan magazine representations of these sectors as well, recalling the trope of woman-as-machine, used as an emblem of modernity in films such as Fritz Lang's *Metropolis* (1927).[84]

Beading and embroidery, though not typically involving machines, can also be categorized as light manufacturing because of their characteristic repetition and detail and the fact that similar jobs were some of the lowest-wage feminized work in the garment industry. That they involved the same repetition and tedium as machine work but had to be done by hand made these jobs even less desirable than machine sewing. In at least one studio, immigrant women carried out such jobs. Ronald Davis describes the separateness of the "sixteen Mexican women" who did all of MGM's embroidery by quoting studio publicist Ann Straus, who said that they brought brown bag lunches and never left the second floor of the wardrobe

Fig. 14. MGM women's wardrobe department circa 1950. (Courtesy of the Academy of Motion Picture Arts and Sciences.)

department: "They just sat and beaded and embroidered all these beautiful costumes that were designed by the MGM designers," never knowing the people whose clothes they worked on for a given film, but only that "it was production number 1420 and that was all."[85] Beth Day similarly states that in the same wardrobe department, "fine seamstresses from Mexico and Japan and Puerto Rico made exquisite hand embroideries and decorations," including one costume for Garbo that "took eight needlewomen from Guadalajara nine weeks to make."[86]

The studio tours pass through other women's light manufacturing sectors such as film laboratories, where women equal or outnumber men onscreen. This is hardly a surprise, given that women were the main workforce in photographic manufacturing before the advent of motion pictures—in photo-finishing laboratories, photographic plate manufactures, and drying, cutting, and retouching film—and assumed similar work in the early motion picture industry.[87] However, not only the presence of female lab workers but also the manner in which they are depicted is significant. In all of the films in which the laboratory processes are explained in any depth, the women are shown in groups of seven to ten, performing near-identical, repetitive tasks.

Women lab workers were similarly featured in the trade reports on growing studios discussed in chapter 1. Indeed, rather than merely

Fig. 15. Men's work versus women's work as represented in pictures of various wardrobe departments (undated). (Courtesy of the Academy of Motion Picture Arts and Sciences.)

implying the presence of women workers through descriptions of facilities or workflow, as was the case with other feminized sectors, the gender of the employees was frequently highlighted in displays of laboratory assets— evidence of modern efficiency. For example, a feature published on the Selig Company's growth and development described "rooms of girls" in the lab in 1909; another, in 1911, explained that "deft-fingered girls" were required in order to carefully watch negative printing for errors; and a third, in 1915, displayed the laboratory that employed "hundreds and hundreds of girl experienced operators."[88] At Paramount in the 1920s, "50 men and girls" were hired and, in many cases, trained for their work in the laboratory, where prints were "projected on a tiny screen by a girl operator" who inspected it, "marking all flaws, scratches, cloudy or foggy pictures, etc."[89]

The girl operators were referenced frequently in a sector filled with prized studio technology, which might otherwise have been assumed to be a male sanctuary given the technological competencies required to process film. This mode of filmic and photographic representation was likely a deliberate choice on the part of the studios or journalists (surely there were moments when the women were standing, looking up, or assuming different postures from their neighbors). In contrast, male lab workers

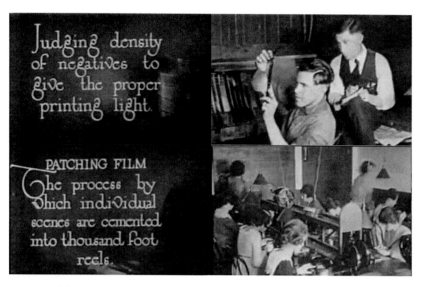

Fig. 16. Men's lab work versus women's lab work as depicted by *A Tour of the Thomas H. Ince Studio* (1924).

were almost always pictured in a variety of poses (standing, sitting, walking between tasks) even when they all seemed to be doing the same job. In the studio tours, male workers appear engaged in multi-staged, active, and self-directed work.

Stages of the process involving heavy machinery (large drying racks and bins, mechanized baths, and so on), individual judgment (timing of exposure, negative density, and so on), and technological knowledge (developing and transfer machines) were assigned to men, along with supervisory positions over both male and female work sectors, as seen in the Ince tour. It fell to women to cut negatives according to the editor's specifications and to assemble prints by patching the different shots together with the small splicers and film cement. Women also inspected film for defects, again using light, simple tools for tasks that required only a few movements of their "deft" fingers. The real impetus for women's employment in these areas was more likely related to the bottom line: early female lab workers received $7 to $12 a week, whereas men received $2 to $5 more.[90]

The studio tours films present only two possible exceptions to the rule of women in the light manufacturing sectors of laboratories and editorial departments. A woman is visible in the group shot of the "cutters" on the MGM *1925 Studio Tour*, and a female worker (who may or may not be an assistant) appears in a shot of the Fox editorial department, sitting behind and deeper in the frame than the male editor and supervisor working there. Many early cutters had been women, but that changed with the growth of the studio system, accelerating after the advent of sound when, as Edward Dmytryk recalled speaking of his time at RKO, "There was a stream of personnel flowing into the cutting department, and because of the real and imagined difficulties involved in cutting sound, that stream was almost exclusively male." Dmytryk proceeds to explain how the shift took place: "Some of the women, intimidated by the new complexities, voluntarily retired. Others were replaced by department heads who felt that the ladies couldn't cope; that at least some of these ladies remained top editors right through the transition had little effect on their prejudices."[91] Film editing and its status as a "female-friendly" field are discussed in greater depth in chapter 5. For now, it is worth noting that female editors were less rare at studios in the 1910s and 1920s than female workers on set or in most craft or technical production departments, but this is not reflected in the tours' display of a mere two women who might possibly be editors. Otherwise, the combined forty-plus female lab workers on display in the studio tours films are relegated to light manufacturing. Photographic work of this sort had been gendered-female prior to the advent of motion pictures, but the feminization of new areas of work specific to the film industry, such as patching and joining, indicates that that logic was not merely imported with old jobs at new studios but extended to new jobs as well.

The logic of feminization was similarly extended to an area of animation labor that grew significantly with the introduction of animated feature-length films. Though the upper levels of the animation process continued to be reserved for men, women began to be hired at Disney in the 1930s because of the large number of workers needed in other sectors of animated feature production for the making of *Snow White* (1937). As artist Mary Ford was told after submitting work for consideration to the Disney animation training school, women did not do "any of the creative work in connection with preparing the cartoons for the screen," and "the only work open to women consist[ed] of tracing the characters on clear celluloid sheets with India ink and filling in the tracings on the reverse side with paint according to directions."[92] Inking and painting, as these jobs were known, required more skill and initial training than most light manufacturing jobs. However, many aspects of the work and the conditions in what in the mid-1930s became an all-female ink and paint department at Disney closely link it to women's light manufacturing. In a curious reversal of the correlation of women with paper planning and men with physical production described in chapter 1, Walt Disney's "girls," as he called them, were hired to defray the high cost of producing animated features by carrying out the *actual* production, applying the paint that would show up onscreen to animation cels that had only been penciled in or planned by animators.[93]

Though technically it was work on the finished product—the film itself—inking and painting was more akin to a factory job than a crew position. Ink and paint work correlated with light manufacturing in the typical separation of female from male workspaces that Disney concretized (with separate buildings) after the company's move to its Burbank studios and increasingly underscored by a production culture in which heterosociality was discouraged. Eventually the ink and paint department became known as the Nunnery. Inkers and painters worked in low light and were monitored for speed and accuracy, and pushed to produce eight to ten cels per hour with only fifteen-minute breaks. These women sometimes worked more than eighty hours a week to keep films on schedule. They were compensated with low pay and low status compared to their male peers, receiving $18 per week in 1941 compared to some animators' salaries of $300 per week.[94]

Other studios saw the benefits of feminized ink and paint sectors and followed suit with their animation divisions, where ink and paint departments were sometimes referred to as "hen houses."[95] The MGM cartoon division grew from a staff of twenty-five to eighty-six members between August and November 1937, just a month before the premiere of *Snow White*, and many of the new hires seemed to be women. As one cartooning employee exclaimed in the studio newsletter, "Speaking of GIRLS! We wonder who is responsible. Never before so many 'lovelies' in just one small department."[96] In her book *Living Life inside the Lines: Tales from the Golden Age of Animation*, former ink and paint worker Martha Sigall describes an all-female ink and paint staff working on Warner's *Looney Toons and Merry Melodies* cartoons as well.[97] When she was first hired to work on Warner cartoons in 1936, Sigall's apprentice painter salary was $12.75 for a forty-hour week.[98]

The boundary between women's and men's animation work was maintained via the normative gender values that placed women's eventual marriage above their careers. Disney ink and paint workers seeking advancement were told that whenever a woman had became skilled enough to animate in the past, she would quit to get married, and that for this reason Walt Disney was "thumbs down on promoting girls."[99] This rationale held that the investment of time would be wasted on women whose true calling lay outside the workroom. Really, the policy served management's interests, sustaining its source of cheap, feminized labor by restricting women's workplace mobility. As the Disney animation rejection letter stated, the possibility of advancement beyond the feminized sector of ink and paint remained nonexistent for most women until World War II, when a few were promoted. The earlier, sex-segregated conditions largely resumed after the war; Mervyn LeRoy reported that though the field was open to women, "there are very few women animators," adding that "women are used exclusively for inking and painting, which require patience and skill."[100] This gendered division of labor remained until at late as the 1980s. Animator and historian Tom Sito reports that, by 1989, ink and paint made up the largest category of workers at animation studios, and was still almost entirely made up of "women who were held back from moving into the ranks of animator, layout, or direction."[101] Had these women been eligible for promotion or higher wages, their numbers would have threatened not only the masculinized power structure but also the very cost control they were hired to ensure.

Women's Professions

Lack of advancement prospects was a common characteristic among all women's jobs to some degree or other. This was even true of teaching and nursing, the most desirable of feminized sectors elsewhere in American society, which, because of their training requirements and elevated status and pay (relative to other women's work), can be classified as women's professions. Women's employment in these fields was culturally acceptable because they were tied to women's domestic roles as mothers and wives, which had in the past required them to act as midwives, healers, nurturers, and teachers. For that reason, it was thought that teaching and nursing "did not require sacrifice of the feminine role" as other professions did, and such roles were primarily associated with women by the late 1800s.[102] These professions also required elements of service and emotional labor, delivered not just laterally to equals and upward to "betters" (as a lawyer serves clients) but downward, to some of the weakest members of society. This characteristic ties women's professions to other types of women's work, and indeed, the 1913 women's vocational manual cited earlier lists nursing, which was still gaining respectability because it required little formal training before the Civil War, under Domestic and Personal Service.[103]

Studios included some traditionally domestic workspaces for the benefit of studio workers. Although spaces in production such as the camera department were marked, through their heavy machinery and male technicians, as the realm of men, other spaces such as hospitals and schools were designated the realm of women. This arrangement enhanced the sense of studios as simulations of the outside world, complete with masculine and feminine spheres.

A woman identified in a caption as "Nurse Peterson" appears in the 1925 Studio Tour, standing in front of the MGM hospital, referred to elsewhere as the first aid department. In 1930, a Screenland feature on women behind the scenes at MGM reports that Peggy Coleman "is chief nurse for this studio," and that she "directs the studio hospital and looks after the ailments of the famous folk of films." By 1940, that studio's hospital was open around the clock because of night shooting and was staffed by a female doctor, Dr. Helen Jones, and three nurses. Jones reported in her own studio newsletter column that the staff saw an average of 150 human patients per day, along with the occasional chimpanzee.[104] Though MGM by some accounts had a doctor, a dentist, and even a chiropractor in residence, most studios did not keep on staff full time.[105] Instead, nurses typically staffed on-site hospitals or clinics and stood by in production to lend first aid if needed.[106]

Though dance instructors are the only teachers on display in the studio tours (at MGM and Fox), many others were present at studios. Child actors typically took classes at studios' schoolhouses (if they had them) or worked with tutors on set.[107] Perhaps most well known was MGM's Little Red Schoolhouse, a Spanish Revival building, where Mary McDonald served as head instructor for years. For pupils who attended classes in the mornings and met with tutors on set in the afternoon, McDonald functioned as a principal, directing teachers who came from the Los Angeles public school system to teach the studio's children.[108] Ida Koverman, Louis B. Mayer's secretary, took great interest in the lives of children at the studio, and reported in the studio newsletter in 1936 that MGM had taken additional steps on behalf of its younger contract players. On Mayer's authority (and, seemingly, Koverman's initiative), a recreational area was built for the children in a bungalow that came to be known as "The Little Green Room," which Koverman wrote about in the December 1941 issue of the *MGM Studio Club News*. She explained that the Green Room's supervisor was "studio mother" Caroline "Muzzy" McPhail, who watched over all children and adolescents and kept "record of their whereabouts at all times." Koverman added that McPhail also sat with the children at their table in the commissary, trying her best "to keep them from eating rich pastries, hot dogs and milk, pickles and ice cream, etc., but I'm afraid in many instances without much success." The Green Room, staffed by "various tutors and coaches employed by the studio," hosted not only tea and birthday parties but also distinguished visitors. Studio talent scout (and later casting director) Bill Grady, it was reported, "drops in on the bungalow family very often to see how his 'chicks' are coming along, and give them good advice to help them on their way to success and stardom." Pictures of McPhail teaching a group of children accompany Koverman's article, underscoring the Little Green Room's representation of traditional notions of home and family as the feminine sphere.[109]

Though framed in less familial terms, Bill Grady and other executives associated with casting and talent departments used teachers and coaches

The Little Green Room By Ida R. Koverman

For several years, ever since Judy Garland and Mickey Rooney made the M-G-M lot their home, the question of a suitable place for our young players to spend their time between school hours and studio work, has

The Author
News Staff Star

been the subject of discussion, until about a year ago, Mr. L. B. Mayer solved the problem by arranging to install a sort of recreation rest room in one of the small bungalows with a studio mother in charge. Here the children and young people were to report and a record of their whereabouts at all times to be kept. Mrs. Caroline McPhail was placed in charge and the "Little Green Room" came into being, with all sorts of games available for the little ones, a piano, radio, books, and best of all, a tiny electric grill and tea service.

The birthday of any of the frequenters of the Little Green Room is never forgotten and many a merry little party takes place with "Muzzy" McPhail presiding over the tea pot.

One of the parties planned by "Muzzy" McPhail and Virginia Weidler was a "candy pull" and turned out so disastrously that from then on, candy pulls were taboo—it

was a most "sticking" affair and took some time to get everything back to normal in the bungalow.

The Little Green Room has had many distinguished visitors, among them the famous conductor of the New York Symphony Orchestra, John Barbirolli, and his wife; the well-known music critic of the New York Times, Olin Downes; High Commissioner of Australia, Sir Ronald and Lady Cross and their two daughters, Diana and Angela; Dr. James Wood, president of Stephens College of Columbia, Missouri; Mrs. Mabel Walker Willebrandt, of Washington, D. C.; Miss Amparo Iturbi, concert

Caroline McPhail pianist and sister of Jose Iturbi; Sir Victor Sassoon, reputed to be the richest man in the world; Elsie Janis; the famous blind pianist, Alec Templeton, and Mrs. Templeton; Mrs. Cornelia Vanderbilt; Lady Furness; Miss Dorothy James, daughter of Governor James of Pennsylvania; Mrs. William F.

ior Orchestra, and many others.

The Little Green Room is the pride and joy of our famous talent scout, General William Grady, known to his intimate friends as "Bill"—he drops in on the bungalow family very often to see how his "chicks" are coming along and give them good advice to help them on their way to success and stardom.

Mickey Rooney is a frequent visitor and always entertains the

Defense Bonds

Are Like

Money

In The

Bank

OLD GLORY *waves over the little Green Room where "Muzzy" McPhail instructs the youthful Metro-Goldwyn-Mayer players in the triumphant history of their Flag. The children* are, left to right: Darryl Hickman, Darla Hood, Virginia Weidler, Anne Rooney, Larry Nunn, Richard Haydel, Jackie Horner, Bobby Kelly, Gene Eckman.

Pouch, Director General of the D. A. R.; Miss Frances Marion, famous screen writer; Miss Anga Enters, internationally known dance pantomimist; Peter Meremblum, conductor of the famous Meremblum Jun-

youngsters at the piano, and the chief executives of the studio drop in quite often for informal visits. Here, too, may be found the various tutors and coaches employed by the
(Continued on Next Page)

Fig. 17. The Little Green Room and its inhabitants as described by Ida R. Koverman and depicted in the December 1941 *MGM Studio News*. (Courtesy of the Academy of Motion Picture Arts and Sciences.)

to nurture their adult "chicks" as well. Newly discovered actors under contract were subject to a process of grooming and training, an effort carried out by a fleet of speech and diction coaches, singing and dancing teachers, and other teachers of anything a motion picture career might call for, including etiquette, riding, swimming, and language.[110] Talent-grooming departments at studios were staffed and often led by women, including head drama coach Lillian Burns at MGM, Alice Kelly and Helena Sorrell at Fox, Zee Silvonia and later Phyllis Loughton and Charlotte Clary at Paramount, Malvina Dunn and Sophie Rosenstein at Warner Bros., and Benno Schneider at RKO and Columbia.[111] Some talent workers, along with workers in publicity, were also responsible for organizing stars' private lives to fit their public personas, because featured players "were expected to drive expensive cars, live in impressive houses with servants, and dress stylishly."[112] Though there were many male teachers and coaches, talent departments were female friendly and even female dominated, because the characteristics of the work of teaching and grooming tied these efforts more closely to the domestic sphere and to feminized fields.[113]

Clerical Labor

Finally, there was the women's sector of clerical labor, which may on its surface seem to have less connection to the domestic sphere than and less in common with other categories of women's labor, given clerical workers' frequent proximity to high-status, male-dominated sectors (for example, as a secretary in an office of male managers and executives), and the fact that many male studio employees occupied such positions. However, as explained previously, clerical technology (typewriters, stenography machines, and so on) and most clerical roles had been gendered female in the nineteenth century because of their routine processes and requisite attention to detail, aligning clerical machines more with "light" manufacturing machinery than with masculinized technologies or heavy machinery. Further, clerical labor's low status, lack of potential advancement, and service aspects (depending on the job) located it closer to domestic service or women's professions. The efficient reorganization of the office led to a gradual departure of male workers from most clerical roles in the late 1800s, other than those that still had managerial authority and potential for advancement beyond the clerical realm—or involved some more masculinized technology or machinery. These parameters also guided men's participation in clerical sectors at early studios, as evidenced by male workers' presence at the head of clerical departments such as accounting in trade journal discussions of growing companies, as well as their occasional appearances in the tour films.

A male clerk is featured in the 1925 Universal tour, sitting sentinel outside of the studio manager's office, and MGM's five male "reception clerks," presented at the start of its 1925 tour, are shown standing in front of the studio gates rather than seated at desks or typewriters. The shot of the clerks is sandwiched between title cards that state first, "It would take a 'One-Eyed' Connelly to crash this gate!" "Unless these reception clerks had been tipped off to Okay your visit." Positioned as literal gatekeepers with power to grant passage beyond the studio walls, these roles are framed as necessarily masculine. The MGM clerks who manned the front desk just inside the studio entrance were called the "front office boys" in the *MGM Studio Club News*, which profiled each of the young men in 1927 in an effort to "attain a better light of understanding" between them and other studio workers and to make various departments more "front office conscious."[114] The later use of police and security guards as studio gatekeepers corroborates the notion that, as one studio history puts it, "security at studio gates was a serious business."[115] The only other male workers displayed in roles related to clerical labor (all of whom are shown in the Fox tour) are two leaders of administrative departments (the personnel executive and treasurer), whose positions are managerial or executive rather than clerical, and a male clerk shown at the paymaster's window swamped with workers. The latter was another area of male clerical participation (and financial control) reflecting themes of gatekeeperism and male technology.

In his February 1937 "Studio Tours" column for the *Warner Club News*, Fred Pappmeier provides the history of the timekeeping and tabulation departments at Warner Bros., which, in keeping with the pattern outlined thus far, developed from the early film model of clerical and accounting as a minimal and haphazard presence at the studio, ancillary to the production process, with "only one man for 300 people" located "in a corner of the purchasing department."[116] Over time, the department grew along efficient lines, with a large staff of clerical workers undergirding the production process. By the 1930s, thirty-one assistant timekeepers were "dispatched to various departments to keep track of who is working on which project." Female clerical workers in the time office supported the timekeepers and managed a mechanical payroll system for the studio's three thousand employees. Pappmeier's history replicates the earlier transition in other industries from the Victorian model of masculinized clerical work—under which one or a few self-directed male clerks worked as masters of an entire office or department's processes— to the efficiency model of scientific management, in which clerical tasks were separated, mechanized, and deskilled, with the lowliest among them reserved for female workers. In Warner's mechanized system, the male timekeepers used their individual judgment to assess labor use across the studio, while female operators filled in time cards for employees and punched them as workers came to and from work so that they could be run through tabulation's new "brain" machine, which calculated pay, overtime, and so forth every Saturday.[117] This gendered breakdown of labor extended to tabulation department leadership—it was run by a male department head and his male assistant, and male operators maintained the heavier technology of its twelve IBM "brain" machines, leaving to women the lighter, detail work of maintaining the files of time cards and punching holes in them.[118]

The male wire operator shown in the MGM tour is representative of hiring practices at that studio until 1942, when the studio newsletter reported that the "first ever girl" had been hired as an assistant telegraph operator in the telegraph department, with its masculine technology, because of wartime staff shortages.[119] Similarly, mimeograph machines, because they were considered heavy machinery, were run by men in clerical departments such as script and stenography, and male clerks were sometimes assigned to clerical roles in departments with heavy machinery, such as the camera department. By the 1880s, answering telephones and operating switchboards had been feminized in broader American culture according to the rationale that women were better suited to the technology because of their "gentle voices, nimble fingers, and mild tempers" (and because operators were needed in high volume and women could be paid low wages).[120] These jobs were likewise feminized at studios, as well as in the works of popular culture produced there, which frequently featured female operators either onscreen or as disembodied voices heard through a telephone. In 1938, Jean Booth reported in the MGM newsletter that one of eighteen female operators working with her at the switchboard wanted to know "why telephone-operators always chew gum—in pictures. They seldom do while working."[121]

Female clerical workers appear in nearly all of the studio tours, though the Universal and Ince tours merely feature shots of women taking notes on scripts in or around production but do not stipulate their role. Note-takers and "script girls" are visible at Vitagraph in *A Trip Thru a Hollywood Studio*, and related workers are shown at Fox in the research department. By far the biggest clerical presence can be found on the MGM tour, in which a title card explains that, after stories were written, "Miss Underwood, Miss Remington, et al. make copies of stories for all departments," leading in to a shot of an arrayed group of seventeen to eighteen smartly dressed women, presumably making up an early stenography or typing pool.

Whether described explicitly or merely implied (for example, by shots of the large, imposing administration buildings as in the Universal *Behind the Screen* tour and the Fox and MGM films), as the next chapter explains, clerical workers, most of them women, were a growing presence in nearly every department of every studio. By the 1920s, they were a necessity for doing business, as Pandro Berman discovered when he was fired as supervisor of the Columbia cutting department for what he described as his "ignorance of the techniques of running an office." Berman didn't have a secretary, instead writing longhand letters with instructions related to cutting until his supervisor summoned him and asked to see a copy of one such letter that Berman had sent the New York office. When Berman replied that he didn't have a copy, "he threw me out."[122] Thus, the relative lack of female clerical workers visible in the studio tours, given their ubiquity by the 1920s, demonstrates not their absence from the production process but their spatial remove from the movie sets and their lack of perceived importance to the creative and managerial work of moving making.

Women in Studio Tours—Revisited

The following list of women's jobs generated earlier in this chapter from the studio tours has been sorted into the various categories of feminized labor:

Performance
  Actress
  Dancer
  "Mannequin"/Model
Literary/Intellectual
  Reader
  Researcher
  Screenwriter
  Scenario Writer
  Scenario Department
Domestic Service
  Maid/Servant
Domestic Arts/Crafts
  Costume Designer/Dept.
    Head
  Dance Instructor
  Hairdresser
  Modiste (seamstress)
  Modiste Assistant
  Pianist
  Wardrobe Assistant

Wardrobe Mistress
Wardrobe Staff
Domestic/"Light" Manufacturing
  Cutter
  Film Inspector
  Film Lab Worker
  Negative Cutter
  Patcher/Assembler
  Seamstress
Women's Occupations
  Nurse
  Shopper
  Wife/Mother/Sister
Clerical
  Copyist/Typist
  Mail
  Note-Taker
  Script/Continuity (Script Girl)
Exceptions
  Art Director's Assistant
  Publicity Department

Whether women had really been effectively marginalized in all other labor sectors by the time each of these films was made, or were simply not selected for on-camera display when they were present in outlying fields, their professional prospects at growing studios were narrowing from the early days, when they were encouraged to double in brass all over the studio.

Nearly all of the female workers who appear in the films do so under one of the categories of work that had, by the 1920s and 1930s, been incorporated as feminized into most studio labor systems. Women who appear in the film outside of those categories of women's work fit either under the category of performance (actors, singers, dancers) that acceptably aligned with gender norms, or in the female-friendly subset of so-called literary or intellectual professions, many of them tied to clerical labor, with the rest rendered acceptable due to feminine associations of the work (typing, done in private). There were two exceptions to this arrangement, but only one—the art director's assistant shown on the MGM tour—is likely a true exception. As chapter 5 explains, though there had been more female publicists at earlier film companies, in the late 1920s and early 1930s, the field became male dominated along with most others at studios. According to Ronald Davis, the practice of hiring women in publicity did not catch fire at MGM until Howard Strickling's tenure later in the decade.[123] Though it may be impossible to definitively prove which position she held, it seems equally likely that the female worker shown with MGM's publicity department is a clerical worker.

Conclusion

Though increasingly excluded from high-status fields, women were not excluded from film production entirely. Indeed, if individual female movie makers were crucial shapers of the early industry from positions of creative and managerial importance, after feminization, groups of female movie workers were a cornerstone of movies' mass production. As studios swelled to the size of small cities, they needed women's low-wage labor as much as they needed investment capital; both were means to scaling their expanding businesses.

Acknowledging the importance of feminized labor to film history helps to dispel the myth of women's total exclusion from film production during the studio era without diminishing the accomplishments of those rarer female movie workers and makers who successfully transcended such sectors. The next chapter examines one branch of women's labor outlined here—the clerical branch—in greater depth to give a sense of how greatly it impacted the industry, and of the importance of women's labor in general to the larger system of film production in operation at the studios.

3

The Girl Friday and
How She Grew

Female Clerical Workers
and the System

There are ninety of us here, all career-bound in pictures
but in the tadpole stage . . . an odd sprinkling of
stenographers, script girls, assistant cutters, designers,
a librarian or two and one honest-to-goodness writer.

—"Madge Lawrence" in *I Lost My Girlish Laughter* (1938)

In February 1937, the *Los Angeles Times* announced the engagement of Ring Lardner Jr., "screen writer and son of the late novelist," to Silvia Schulman, "secretary to motion-picture producer David O. Selznick."[1] A year later, Schulman published a novel called *I Lost My Girlish Laughter* under the pseudonym Jane Allen. The story follows a bright, hopeful young woman to Hollywood, where she quickly becomes disillusioned by her work as a secretary to a major producer and eventually leaves his employ to marry a colleague. Though a fictionalized version of her experiences, Schulman's book is accurate in many of its depictions of studio-era life, including its chaperoned women's boarding house (a fictional version of the Hollywood Studio Club) with its mixture of studio extras, stenographers, designers, researchers, and writers.

In 1936, *Photoplay's* "They Aren't All Actresses in Hollywood" profiled eight female studio workers, "each representing one of the eight studio trades open to women" outside of acting. "These, then, are your chances in Hollywood," the article prefaced. The occupations were only slightly more varied than those described in Schulman's boardinghouse: seamstress, costume designer, interior decorator, waitress, hairdresser, secretary, writer, and singer.[2] A women's vocational manual from the mid-1930s was slightly more optimistic, including screenwriting and scenario editing among a list of jobs where "quite a group of women" were said be extremely successful, along with editors and some other fields related to scripting, including continuity and dialogue writing, reading, and research. Women's chances seemed even better in wardrobe, as script girls and patchers, and better still in "all the positions common to every business, such as stenographers, typists, and secretaries." Still, the manual cautioned, "There is but one full-fledged woman motion-picture producer in the business, Dorothy Arzner."[3] No mention is made of directing, a fact that starkly contrasted with the 1920 edition of the same manual, in which an entire article is devoted to women's considerable qualifications for and prospects in that field.[4]

Women movie workers numbered in the hundreds at each of the major studios of the 1930s and 1940s, yet their increased presence across a variety of departments did not help them infiltrate the masculinized specializations within those departments. If anything, women's options became more limited as time went on. As studios entered their age of maturity, women were no longer being promoted to the director's chair, and those female directors who remained in masculinized fields became outmoded under the system that was developing and outnumbered by the bulk of female studio workers in feminized work identities more integrated with studio interests. A 1936 MGM studio newsletter listed the names and departments of all workers continuously employed by the studio since its opening in 1924. Roughly a dozen of the 118 names on the list belonged to women, and all of them worked in the feminized sectors of clerical labor (accounting, script, printing, telephone), hairdressing, wardrobe, and negative cutting.[5] By the early 1950s, wrote script supervisor May Wale Brown, "If you were a female who wanted to work on a movie set, but didn't have exceptional beauty or talent, you had five choices," which were character actress, extra, wardrobe, hairdresser, or script supervisor.[6]

Though individual female movie makers' professional prospects shrank under studio efficiency, as a group, female movie workers grew in importance during the studio era in their significance to the system of large-scale film production through which the industry maintained its massive output of roughly 350–400 features per year. Examining how distinct classes of feminized workers functioned within this system reveals just how

seamstress

designer

decorator

waitress

There are two hundred girls waiting for Sally Paige's job sewing tucks. Eli Benneche (right) studied for six years how to dress sets at M-G-M. Alvina Bryan admits that "slinging hash" for stars is t o u g h work. Edith Head's c o s t u m e designing was the happy ending of a practical joke on H o w a r d Greer. All aren't so lucky

They aren't all Actresses IN HOLLYWOOD

THIS story is simply a very frank and truthful answer to the fifty thousand women who write yearly to the film studios asking:

"Is there a job for me in a studio? I have no ambitions to become an actress but I am an experienced secretary (or hairdresser or dressmaker or singer or interior decorator). Pl ase tell me what MY chances are in Hollywood."

And because these letters come from women employed in work ranging from domestic service to gown designing, I have decided to let eight Hollywood working girls (each representing one of the eight studio trades open to women) answer the clamorous plea of these fifty thousand American women.

These, then, are your chances in Hollywood——

IF YOU ARE A SECRETARY: From the case of one Simonne Maes, ten years a studio secretary, you may deduce your own conclusions. A decade ago when she was attending Columbia University and translating French and Spanish books on the side, she too longed to come to Hollywood and work in one of "those glamorous studios."

Well girls, she came, and luckily for her she brought a brand new husband with her. He provided shelter and food during the eight months Simonne hounded studio gatemen for a chance to see just one hiring boss. She filled out during that interlude no less than eighty-five applications.

"That was in 1926," Simonne recalled for me, "and it was comparatively easy to break into the studios. You see, there were only three girls to every secretarial job then, instead of the twenty waiting for every opening today."

She finally started at Universal as a stenographer at twenty-five dollars a week and from that moment everything happened just as she had dreamed it would. She was promoted and raised in salary. After seven years she was earning sixty dollars a week and she had met, personally, such stars as Lon Chaney and Reginald Denny and knew many others by their first names. She worked for all the important directors, writers and executives in the studio and was often called upon

Fig. 18. "They Aren't All Actresses In Hollywood" examines the eight trades open to women. (*Photoplay*, September 1936.)

important their work was to Hollywood's overall project of systematized film production on a massive scale. For that reason, while the previous chapter toured feminized labor at studios, locating branches of women's work on lots and within the industrial logic that organized the studio system, this chapter delves more deeply into one branch: women's clerical labor and, particularly, studio secretaries.

Clerical workers were perhaps the most prominent of the feminized groups on the lot by virtue of their large numbers and their widespread distribution across various studio departments. As the primary producers, distributors, and custodians of studios' clerical output, these workers ensured that production could function on such grand scale. As the designated recorders and rule minders of production, they brought rationality and control to a creative process in which most women could no longer directly participate. Yet, as the second half of the chapter explains, women's work in the classical Hollywood era extended beyond the visible labor they were assigned in these feminized sectors. Outside their gates, studios promoted themselves to exhibitors and the public by enlisting images of such female workers as secretaries and script clerks to represent the power and sovereignty contained within the walls of their glamorous motion picture cities. Inside the isolated studios, a different narrative was unfolding. As studios embraced traditional notions of women's place in business, women were expected to perform femininity and sexuality on the job, fulfilling unspoken duties and serving at once in professional and emotional capacities that, although often at odds, were an expected part of their work.

Amid production's intense working conditions, female employees assumed an emotion-based second shift during long hours and late nights at the office. They received neither compensation nor credit for this work. On the contrary, rather than acknowledging their additional responsibilities, these women's workplace culture required yet another level of gendered performance from them to camouflage the very significance of their own labor.

Perfect Pictures: Paper-Based Efficiency and Clerical Divisions in Full Bloom

By the late 1920s, efficiency flowered out of the first decade of experiments in scientifically managed film production, as studio managers exacted control over the creative process in ways the earlier managers discussed in chapter 1 had only dreamed of when they wrote about their goal of making "perfect pictures" and of engineering scripts (via efficiency) to be directed as written.[7] Richard Koszarski calls the efficient production systems developed by even such pioneers as Thomas Ince "primitive in comparison with those employed later in the silent period by more mature studios such as Paramount and MGM."[8] The latter—one of the largest employers in Southern California by the 1930s—grew from six glass stages and forty-two buildings in 1924 to thirty concrete stages and 177 administrative and support buildings at the height of its development in the 1940s, when, as one executive reflected, "You could get anything done on the lot except be born or buried."[9]

If the front office fixated on efficiency—providing specialized, standardized, rationalized control over production—employees often cursed it. Publicist Ann Straus remembered that efficiency experts, dispatched from MGM's New York offices, required every studio worker to write out a detailed description of his or her job, then undergo an interview, after which "they fired a number of people and put a stapler on everybody's desk. That was to make them more efficient!"[10] Editor and director Robert Parrish recalled 20th Century–Fox's control of assets and personnel extending all the way down to the parking spots:

> When a firing was imminent, the studio management notified the payroll department and the sign-painting department. When these departments were poised and ready to strike, the unlucky employee was notified. On D-day a signal was flashed from the front office (Zanuck? Koenig? Goetz? Schreiber? Somewhere up there), and the departments sprang into action. A final check was sent to the now ex-employee's agent, and as the ex-employee drove off the lot, the sign painter rushed out and blotted his name off the asphalt forever.[11]

Departmentalization and separation of work processes allowed for the management of creativity and subjectivity at a remove from the front office. An ever-expanding class of managers controlled the assets and personnel surrounding creative work and thus isolated creative processes from other labor. In his 1941 study of Hollywood, Leo Rosten described an "immensely ramified division of labor," in which "producers found it necessary to create lieutenants and sublieutenants in the growing army of movie makers—assistant producers and associate producers, supervisors and story supervisors, A producers and B producers."[12]

Attention to detail had been an oft-mentioned prerequisite for efficient production, as outlined by earlier studio managers in 1910s trade articles with titles like "Devotion to Detail at Lasky's," "Attention to Detail Makes for Success," and "Ince Makes War on Inconsistency" that claimed that systematization would relieve directors of the "detail and routine which ordinarily are a handicap against the highest artistic results."[13] The details in question were both on-screen minutia (for example, historical accuracy, continuity), and off-screen production responsibilities not directly related to the purely creative work of direction.[14] These early managers promised to make everything and everyone efficient, except the director, from whose shoulders would be lifted "a thousand and one mechanical details that could do nothing but interrupt him," leaving him free to concentrate on the story.[15]

Viewed from another angle, this managerial attention to detail also relieved the director of much relevance in the chain of command at studios, where previously, as the center of creative authority, directors not only led the crew in production but also had more decisional power over which films they made and how they made them. Director King Vidor summed up the effect of big studio efficiency on the creative process, saying "The multiplication of men and machines behind the camera made it increasingly difficult for the director to retain his individual viewpoint."[16] A 1938 Screen Directors Guild survey reported that studios' producer ranks swelled from 34 producers producing 743 films in 1927 to 220 producers producing 484 pictures by 1937.[17] The studio director, according to then-independent producer David Selznick, had become "a cog in the machine," and was simply handed his next project in script form, "usually a few days before he [went] into production."[18] In a 1939 letter to the *New York Times*, a disgruntled Frank Capra estimated that "80 per cent of directors today shoot scenes exactly as they are told to shoot them without any changes whatsoever, and that 90 percent of them have no voice in the story or in the editing."[19] Writers felt similarly disempowered by the compulsory collaboration studios required of them under the rationale of making full use of multiple writing talents. Many believed that the real reason for team writing, in which scripts passed through many hands for various drafts, was producers' compulsive need for control over the process.

Through multiple systems of record keeping, cost accounting, message circulation, and so forth, studio managers collectively managed movie labor and assets, planned films through executive, story, publicity, and casting departments, and tracked films during the production phase through the circulation of scripts, memos, and paperwork between backlots and soundstages and executive offices. Clerical workers were the means of exacting this executive-managerial control, first by creating and managing this paperwork in departments around the studio (as clerks, stenographers, and so on); second through monitoring on-screen details to ensure pictures' consistency and accuracy (as researchers and continuity-minding script clerks); and finally in their absorption of other details or routine practices around the creative processes of their employers (as secretaries and, occasionally, assistants).

Paper-based efficiency now included hourly weather reports for not just Los Angeles but all over the country. Massive libraries and storehouses held all sorts of assets, from sheet music to backdrops and even plaster molds.[20]

Due in part to the large scale on which they were bought and managed, the overhead cost of maintaining said assets (which included studio lots themselves) had risen steeply since the late 1910s, when Anthony Slide reports that little was said about overhead on budgets, and preproduction costs were relatively minor.[21] Overhead costs accounted for 20 percent of films' production budgets by 1929, and rose ever higher in the 1930s.[22] Vidor glumly described the massive organization and attendant costs in his biography, saying:

> I was a victim of a large studio with two-thousand telephones, a police and fire department, a mail department, camera department, laboratory, carpenter shops, plaster shop, wardrobe department, publicity department, commissary, barber shop, accounting, casting, and on and on, ad infinitum. This jumble all came under an overhead of four or five thousand dollars a day, very little of which actually went into the finished film.[23]

Hortense Powdermaker questioned the validity of overhead figures in her anthropology of Hollywood, saying they were inflated to conceal studios' profits for taxes and to prevent actors and directors from asking for a fairer share.[24]

Inflated or not, overhead costs were substantial. Simply maintaining studio lots and their assets necessitated a massive organization and large numbers of workers. Feminized labor allowed studios to achieve this control at lowered cost, and as clerical divisions arose and expanded, women's workplace identity was increasingly tied to the overhead and planning departments in which so many of them typed and filed.

All This and Sadie, Too: Clerical Distribution across the Lot

The scope and distribution of studio clerical workforces was vast, encompassing departments devoted solely to clerical, cost-accounting, and record-keeping functions, as well as clerical subdivisions within other departments. Though they varied slightly in their structure, organization, and specific titles, the same basic categories of work could be found at each studio, along with the same or similar job titles and departments. The executive or front office branch was the seat of decisional authority and oversaw all the work in the studio's several dozen other departments, which are here divided into three categories based on their primary purposes: overhead, planning, and production.[25] Day-to-day administration—maintaining the ongoing operation of studios and managing studio assets—was carried out by what were variously referred to as overhead, infrastructure, or plant operations departments. Overhead departments encompassed cost-accounting functions such as purchasing, human resources functions such as timekeeping and personnel, plant maintenance functions such as custodial, food and service (that is, hospital, police and fire, and craft service departments), and communications functions such as telephone and mail. Planning departments were devoted to preparing projects for production and eventual release, including script or scenario (with story, reading, research, and writers subdepartments), casting, talent, and publicity. Directors and producers were actively involved in the production processes of films, but they were frequently grouped with planning departments in studio organizational documents because of their key role in the preparation of films and their location within studio hierarchies and on the lot (near executives and script-related departments). Production departments were all those devoted to producing the physical product (films), including art and special effects, camera, machine shop, technical (which included mechanics, laborers, builders, millers, carpenters, and so on), electrical, wardrobe, makeup, property, hairstyling, transportation, sound, editorial, laboratory, and music. The dozens of departments that crowded studio lots by the 1930s spanned all these functions and more, and each contained up to several dozen occupations and subspecializations within them, from tool and saw sharpeners and stone masons to carpenters, cooks, and cashiers; yet nearly all of them had one thing in common: clerical workers. Their labor provided the organization and continuity of management needed to allow all of these departments to function as a system, rather than as a collection of discrete parts.

Overhead Departments

Clerical workers—male and female—were heavily concentrated in overhead departments, which, not surprisingly, were located farthest from production both hierarchically and geographically. Accounting, tabulating, timekeeping, personnel, payroll, inventory, insurance, stores, purchasing, cost, legal, mail and receiving, telephone and telegraph, and stenography or secretarial departments were all primarily clerical departments devoted to processes carried out on paper. It is impossible to know exactly which jobs women held in every department and at every studio. However, evidence from specific departments suggests that overhead clerical labor was typically divided between men and women in much the same way it had been in other industries, based on status, pay, and duties.

According to a women's vocational manual from 1926, at that time, men still dominated such positions as head bookkeeper, accountant, and CPA in accounting departments in many other American businesses because "the average business man has been unwilling to concede that the judgment, initiative, and general background essential, for instance, in an accountant, could be found even in a woman trained in the science of the work." The manual's author, Miriam Simons Leuck, explained that women's infiltration of these positions depended on "the woman and the situation," in a sector that was "only now beginning to admit women."[26] Only government accounting work, where there was less competition from men because of the work's lower salaries and reduced potential for outside business opportunities, was more wide open to women.[27]

Outside of personnel departments, which, like commissaries, were female friendly because of the interpersonal and service requirements of human resource management, the same exclusion could be found in administrative and clerical departments at studios in the 1930s and 1940s. Men typically held managerial, supervisory, or otherwise high-status positions in clerical or cost-accounting departments, while women carried out roles with lower salaries and more routine and standardized duties. There was relative gender integration (depending on the department) in the midlevel roles in between. Messengers, low-wage, entry-level positions at studios, were typically male, but as with the heavy machine operators described in the previous chapter, this was largely because of the physical requirements of the job—lifting and carrying sacks of fan mail and canisters of film around the lot—as well as existing gender norms under which, as director Edward Dmytryk recalled, "no well-brought-up lady would be caught dead carrying a package in public."[28] These norms began to shift during World War II when "messengerettes" filled in for male messengers overseas. A specific illustration of the typical, men-at-the-top, women-at-the-bottom division of overhead clerical labor can be found in the example of Warner Bros. and workers in its Studio Office Employees Guild.

In November 1940, office and clerical employees at Columbia, MGM, Republic, Universal, and Walter Wanger Studios voted in National Labor Relations Board elections to unionize under the Screen Office Employees Guild Local 1391, A.F.L., which would represent them in collective bargaining with the Producers Association. The guild was to include "all office, clerical, secretarial, and accounting employees, office and tabulating machine operators, switchboard operators and messengers employed by each of the companies, casting employees whose work is of a clerical nature included in the units of those Companies who employ casting employees."[29] In the same decision, 20th Century–Fox was ordered to hold an election to determine whether its employees would be represented by the Screen Office Employees Guild or by the 20th Century–Fox Studio Office Employees Guild, an unaffiliated labor organization for office and clerical employees at that studio that had represented clerical units in collective bargaining since 1937 (employees at Fox continued to be represented by the Office Employees Guild until another vote was ordered in 1944).[30] At Paramount and Warner Bros., similar self-styled guilds had operated since 1937 for the purposes of collective bargaining between the studios and their clerical workers. The Warner Bros. Studio Office Employees Guild (hereafter, WBSOEG) continued to operate at that studio into the 1950s and 1960s, where groups of guild representatives—elected officers from among the studios' office workers—adjudicated member concerns.

The WBSOEG was well established by the late 1940s and produced abundant paperwork in its processes of defining job duties and determining pay grades and wages for various employees. As a result, today, the USC Warner Archive holds what is perhaps the most complete surviving record of the gender breakdown within administrative and office jobs at studios.[31] WBSOEG records corroborate comprehensive descriptions of the Warner organization published a decade earlier in that studio's house organ.[32] Because they fall quite late in the studio era, mostly concentrated in the mid- to late 1940s, these records give a sense of the scope of the studio's infrastructure at the height of its development and paint a detailed picture of the full proliferation of clerical jobs and their gendered division of labor in a studio permeated by paperwork.

A WBSOEG 1949 ledger of Job Titles and Classifications listed all office-related positions at the studio in November of that year, based not on workers' departments, but on their inclusion in the studio's branch of the Office Employees Guild. The ledger also ranked the positions in terms of pay grade and adjudicated pay raises based on job title and requirements.[33] In terms of sheer numbers, Warner overhead departments tended toward female-dominated clerical labor. Although most departments were at least gender integrated, many departments, such as research, were female dominated in terms of sheer numbers (seven women, two men), the ledger generally evinces a traditional division of labor between men and women along lines of status, technology, and gender norms.[34] In the accounting department, which housed more office workers than almost any other, the breakdown by gender was nearly equal, with twelve men and thirteen women on the payroll. However, more men occupied upper-level supervisory positions (head clerk, senior clerk, and so on), and men held nine of the top thirteen highest-paid positions, with pay grades of A–E where K was the lowest possible grade. Most of the women in this department were secretaries, mid- to junior-level clerks, and IBM punch card operators. In the nearby insurance and purchasing departments, gender breakdowns followed the same pattern, with men holding more senior-level positions and women holding more of the secretarial or junior-level accountant and clerk positions.[35]

Women dominated departments without senior-level or high-status positions such as head accountant. For example, in the business office (which housed a number of office machines), the eleven clerical workers were broken into female telephone operators (eight) and male heavy machine operators (three). The legal department's clerical workforce was all female because it was comprised entirely of secretaries. According to the same ledger, the transportation, tabulation, and telegraph departments were male dominated, likely because of their heavy machinery and masculinized technology requirements, with only a few female clerks and operators of light machinery. Men also dominated clerical jobs in departments requiring lifting and carrying, such as the mail department (all male, both clerks and the aforementioned messengers), the receiving department, and the stores department (which had nine clerks and "checkers" who inventoried lumber and other materials).[36]

Not all studios left behind such complete records as to provide this sort of "snapshot" of labor at a particular moment in time, but descriptions of these departments in studio paperwork, newsletters, memoirs, photographs, and other documents from roughly the same time evince this division of labor. To name just a few examples, the basic processes at the MGM purchasing and

Fig. 19. MGM accounting department, circa 1950. (Courtesy of the Academy of Motion Picture Arts and Sciences.)

accounting departments–those involving the typing of purchase orders and other documents—are carried out almost entirely by women based on photographs of the department.[37] Per organizational flow charts from the 1940s, the 20th Century–Fox legal department's staff consisted of "3 attorneys and 8 girls." Workflows for other departments at the same studio don't specify as to the gender of clerical workers. But in the legal department, where the only clerical roles were as stenographers or secretaries—roles so associated with women that employers often referred to their secretaries as their "girls"—it is assumed that the underlings will be women.[38]

Planning Departments

The other areas of high clerical concentration at studios were departments related to planning, where female clerical workers frequently outnumbered the male executives, directors, producers, and writers they served. Like planning departments in other industries, these departments did not produce a finished product, but rather the intermediate product of clerical output, which guided the actual product through its production and sale. Two such departments—reading and research—were typically led by women and were female dominated at many studios. However, these roles were not strictly clerical, and thus are examined further in chapters 5 as women's film development professions. Most other women who worked in planning departments did so as secretaries or stenographers. Indeed, so many typists were needed to produce and circulate scripts that at many studios, stenographic pools were called script departments.

When given her choice of entry-level work at Famous Players–Lasky in 1919, a young Dorothy Arzner chose a job typing scripts, believing it to be the best place to learn the business, because "all the departments, including the director's, were grounded in script." She was introduced to Ruby Miller, "typing department head," and given the next job that opened up there, at a salary of $15 per week.[39]

In an *RKO Studio Club News* interview, longtime RKO secretary Helen Gregg described her early tenure when the studio first opened, saying "the stenographic department was Betty Roberts, who also doubled in brass as the story department, reading department and personnel department in her spare time."[40] Screenplays were handled by one of a few women since, as Gregg explained, "Whenever a script came in, all of us secretaries dropped whatever we were doing and typed it."[41] The studio's clerical needs would soon exceed such a small staff. Stenographic expanded over time, and by 1941, when supervisor Wynne Haslam described the work there, the department had grown to employ between forty-eight and sixty-five secretaries depending on the season, and supplied all labor for typing, proofreading, and assembling scripts, as well as mimeographing departmental forms and legal agreements and issued a catalog of company-owned stories twice per year.[42]

Before she moved to RKO, Haslam started Warner's first stenographic department when that studio moved to the First National lot in 1928.[43] Sadie Freyer, who came to the studio as Henry Blanke's secretary, was put in charge of the department in the mid-1930s.[44] A *Warner Club News* cover story entitled "All This and Sadie, Too," explained that stenographic's "main concern" was typing scripts "that come in that have to be changed and gotten out right away, which, incidentally, is always." By 1940, Freyer was keeping "an eye on the careers of some seventy-five girls," who were assigned as secretaries across the lots.[45] According to a report by the typewriter maintenance department in the same newsletter earlier that year, there were 439 typewriters in use at the studio, many of them by Freyer's secretaries.[46]

When former Goldwyn secretary Valeria Beletti came to work at MGM in 1927, she described a loose hiring system under which producers or executives often engaged their own secretaries, and the story department brought in temporary stenographers to take dictation from its writers when extra were needed, hiring them on permanently when space became available.[47] A year later, Beletti found herself temporarily in charge of the MGM stenographic department, where a more formal system for distributing workers had taken shape. She explained, "All the men on the lot who want work done have to call me and I had to hire and fire girls according to our needs."[48] By the 1930s, according to historian Ronald Davis, MGM employed "some 125 secretaries, not counting the typists' pool."[49] Mervyn LeRoy put the number still higher at MGM by 1953, when he wrote, "There are about three hundred girls in the department, of whom about eighty are in the stenographic pool."[50] Still, not all writers had their own secretaries. Rather, "If a 'rush' script came through, secretaries across the lot discovered that before they could punch out for the evening, they were expected to type three pages of script. Those pages would be rushed to the mimeograph room so that copies could be distributed to the set the next morning."[51]

Samuel Marx included MGM's script department in the list of five he oversaw as story editor in the 1930s (the other four were the reading, research, story, and writers' departments), though it was Edith Farrell who ran the department for him, supervising "her girls," who "were tested note-takers and speedy typists."[52] A large number of typists were necessary, especially when a script was put into production, at which point the production manager routed "25 copies to the various studio departments to cast, design and build sets, and promote the new production."[53] Located in the writer's building, the department also archived old material. LeRoy called the stenographic filing department "one of the most amazing" in the studio, explaining that, at roughly the size of a small-town library, "It contains all of the material Metro owns, purchased and produced," with a foolproof filing system so accurate that when the director sent for a copy of a film he made years earlier, "It was on my desk in ten minutes."

At other studios, stenographers likewise served script, either working within the department proper or assigned to other script-related departments. Secretary and script supervisor Meta Carpenter recalled that when she first worked at Fox in the 1930s, secretaries were farmed out to new writers to take dictation, and finished scripts were sent to the secretarial pool for retyping and mimeograph copying.[54] At that studio, per internal workflow diagrams from the 1940s, sixty or so secretaries from script might be floating around the lot at a given time in the offices of writers, directors, and producers, but the department's payroll also included several stenography rooms with ten or more stenographers apiece.[55]

Many secretaries came to studios with some previous training, obtained from secretarial schools, stenography classes, or through training books and home practice. In addition to the preparation they already had, new secretaries were often vetted and trained on the job. The *Warner Club News* stenographic cover story frames its explanation of how the department works in dialogue between two fictional employees, one experienced and the other "a new girl." In the following passage, the experienced employee, "Joan," explains how the system works:

> The department acts like sort of a date bureau for the whole studio. When a new executive comes on the lot (but that's not often), or a writer or such, he naturally wants to meet his girl Friday as soon as possible, so his first telephone call is to Sadie Freyer who ably equips his office with its first requisite. . . . One morning when you come in all resigned to tap out another script you'll be informed that you're to go over somewhere and relieve somebody in so-and-so's office, so off you'll patter, notebook in hand, not knowing who so-and-so is, or what on earth he does. . . . After you've passed all the tests and got to know who everybody is and the workings of each department you will eventually be assigned to a more permanent job. There's a host of writers who always need little helpers like you and the other girls.[56]

As RKO's Wynne Haslam explained, a similar modus operandi was in place at that studio, with secretaries assigned to different offices until they found the right fit: "Girls come in on a temporary basis. The department is really like a training school. After they are here awhile (seniority rules when possible), they go out on office assignments. Their work and personalities help make or break the girl."[57] Outside of the stenographic department proper, secretaries and stenographers were present in largest numbers in

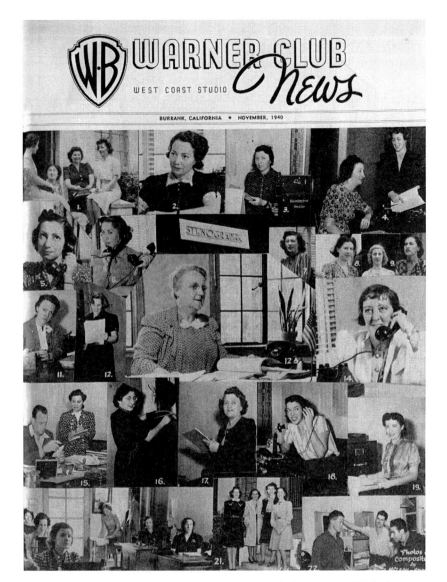

Fig. 20. November 1940 *Warner Club News* cover image, which accompanied "All This and Sadie, Too," a feature on the studio's stenography department. (Courtesy of the Academy of Motion Picture Arts and Sciences.)

departments related to story and other areas of planning. A large proportion of secretaries were assigned to those with most power and creative input on the lot: the movie makers. Executives and producers generally had at least one secretary each, as did directors and some senior writers.

At Warner Bros., the writer's building was called "the Stenographic Department's best customer," housing the offices of important Warner writers and "wards" for other, less-established ones, with another large group of secretaries permanently assigned to "Supervisors and Producers," and finally directors.[58] In the *RKO Studio Club News*, different groups of secretaries wrote columns from buildings in which the largest numbers of them worked, indicating a similar distribution at that studio, with the lion's share assigned to buildings C, B, and G, which housed a mixture of writers, producers, executives, and directors.[59]

Secretaries to these movie makers shared many clearly defined, official duties. Marion Snell, an RKO secretary, gives a colorful description of a typical day of such work:

> Climb the stairs—fumble the key out of the purse and into the lock—throw open the windows and let in some fresh air. . . . But duty calls, and I open the mail. "My day" has begun. And for every other secretary: "Don't forget to send that memo." No, Sir. "Be sure to make that phone call." Yes, Sir. Phones ringing, rushes room 3: Where did I file that @@%!!? letter? . . . "Sorry Mr. Whoozit, he's busy right now—may I take a message?"[60]

A *Harper's* profile of Jerry Wald similarly observes that one of the Warner Bros. producer's two secretaries, Mary Elliott and Lillian Berger, is always in the office by 8:30 A.M. to begin taking down his dictated "barrage of notes, phone calls, and memos," sometimes before she has a chance to remove her hat.[61] A Warner personnel form lists comparable duties for a producer's secretary:

> Takes dictation in shorthand of business and personal correspondence, story outlines and treatments, script changes and continuity, conference notes and interviews, from the producer: transcribes dictated material on a typewriter, setting up in proper format: types copies of scripts, schedules, reports and various miscellaneous data; maintains office files, records and reports; acts as office receptionist and performs routine office and personnel duties for the Producer as requested.[62]

Clearly, though many of their duties were similar to those of secretaries in other industries, the studio's secretary's role also required familiarity with film development and production practices as well as current projects and players. Writers' and directors' secretaries had their own requisite competencies and job-specific knowledge.

Studio-era planning took up to half of the total time spent on each film from start to finish, while the actual shooting took a mere sixth. Preliminary planning took the form of memos and meeting notes that circulated among executives, producers, directors, and department heads. Studio messengers distributed interoffice memos and scripts along with outside mail in deliveries to various departments across the lot, multiple times a day.[63] Though David O. Selznick's memos were legendary, "composed during every waking hour, in his office, his projection room, his car, his bedroom—even the bathroom" and received "by the yard," many other executives and producers had reputations as prolific memo writers and conference holders. At Fox, transcriptions of Darryl Zanuck's story meetings "were bound and sent around to everyone who had attended," along with additional clarifying memos from Zanuck, prompting one Fox director to say that "if everything said about a picture from the time it was begun until it was finished was put in book form it would be far longer than a full-length novel."[64]

Later in the planning phase, meetings were held and budgets set according executives and department heads, who determined

> what the studio had available that could be used in this production and what must be created new. . . . Every man present—from Casting, Wardrobe, Art, Location, the superintendents of construction and electricity, the recording engineer and the cost accountant—had all studied the script and arrived at a rough estimate of what the preparation handled by their department would amount to, so that an over-all estimate of production cost could be reached. "Controllable items" were discussed, and some eliminated, some added. . . . Each man went back to his department to make a revised statement, which he then sent to the estimating department. These detailed estimates were turned over to the production department, which kept the day-to-day records as the production progressed and served as a conscience to all the creative folk to stay in line with the budget.[65]

Clearly, this process of managing and distributing assets created enormous amounts of paperwork. Casting processes were carried out almost entirely through the circulation of lists, memos, and other paperwork, breaking screenplays down by character and discussing which available contract actors (the "Controllable Items" for this part of the process) might play them and whether any outside actors would need to be hired.[66] Publicity departments, typically run like newspaper offices, produced a final product for public consumption: newspaper and magazine stories. Planning—of how pictures would be sold—was nonetheless the departments' main purpose, and their processes were similarly paper based, not only in terms of the stories released to the press but also in the communication about them that circulated between publicists and executives. Fan mail departments were often located in or around publicity, as were still photography departments, where large, typically female staffs filed pictures or signed them and stuffed them into envelopes to return fans' correspondences.[67] Along with the legal department, publicity employed some of the largest clerical staffs on the lot. Paramount, for example, had "twenty publicists plus secretaries" in the 1930s.[68] Later still in the planning process, studio workers prepared to market and sell the film through advertisements and trailers based on the ever-circulating communication from executives, as well as market research in the form of reaction cards filled out by audiences at previews.[69]

This elaborate atlas of information is borne out in the WBSOEG records. At Warner Bros. in 1949, though thirty-seven secretaries were assigned to various producers and writers, publicity had the most in-house clerical positions of any department except the stenographic and mail departments. Women held seventeen of twenty-two clerical positions in publicity.[70] These mostly female underlings would have carried out much of that department's labor (that of typing words on paper), while clerical work in casting was evenly split between three men and three women in a more modest-sized staff.

Production Departments

Though physical production's craft, art, camera, and technical departments had the lowest concentration of clerical workers, there was nonetheless a clerical presence in nearly every such department at studios, in large part because of the studio practice of meticulously tracking projects via scripts, production notes, and other clerical output. In addition to time-keepers, messengers, and other workers who traveled to and from production departments, most production sectors were assigned at least one and as many as ten clerks or secretaries for typing, record keeping, and other paper processes. Together, these workers functioned as an important part of the clerical structure that undergirded the studio system as a whole, connecting production departments to the rest of the studio through paperwork. While such clerks tracked projects through production departments, continuity workers or script clerks (discussed at length in chapter 5) tracked them on sets, taking detailed notes about what was shot, line by line and page by page. These notes, typed and circulated at the end of each workday, connected directors to their editors, as well as to the producers and executives following along through daily rushes.

Clerical jobs in production's craft and technical sectors tended to be more evenly split between men and women because of the requirements of certain departments. Various job analysis forms from the Warner Bros. personnel files include noteworthy stipulations as to the gender of clerks.

The studio's makeup department and "special photographic effects & matte department" both stated that a male worker was required for their clerk positions. In both cases, heavy lifting and carrying are also listed among the job requirements (delivering make-up equipment, stocking the storeroom, and delivery and pickup of film canisters).[71]

Even with such stipulations, enough women still occupied these roles in production departments for the clerical work there to be identified with women. And the most visible clerical role in the production process and the one that was closest to the actual shooting—continuity or script clerk— became so identified with women that the role was more commonly referred to as the "script girl." Still, the fact that production departments typically had the lowest female-to-male ratio of clerical workers is significant given the (rather obvious) fact that said departments were located closest to the spaces of film production, both hierarchically and geographically, and were the places where a woman might have received technical training, gained experience on set, and, hypothetically, become eligible for promotion to crew positions.

Give Consideration to the Young Men: (Female) Clerical Workers as System

Though coworkers with higher creative, managerial, or crew status may have seen secretaries and clerks as mere cogs in the machine, where the studio system was concerned, without cogs there was no machine. No other type of worker could be found in every single part of the studio, from the commissary to the film laboratory. Perhaps because of the mundaneness and low status of their work, the same clerical workers were underrepresented in studio self-representations like the studio tours films of the previous chapter, with the exception of studio secretaries, whose proximity to movie makers and perceived glamour meant they were profiled more often. And though a woman might have been promoted from secretary to casting or publicity assistant by the 1940s, in production departments such as camera, transportation, lighting, sound engineering, and so on, women were practically never promoted out of clerical roles. The production system that functioned by virtue of their labor was all but closed to them. Meanwhile, male clerks, messengers, and "office boys" were well positioned to step into such roles. This much is clear from the "origins stories" of famous directors and other notable figures in production. One particularly pronounced example can be found in Mike Steen's *Hollywood Speaks: An Oral History*. Of the male workers interviewed about their careers, the representatives from the three most elite production fields describe getting started through entry-level clerical work, including producer

Pandro Berman (script clerk), director William Wellman (messenger boy), and production manager James Pratt (timekeeper).[72] In 1938, the *MGM Studio Club News* reprinted an editorial that reportedly first appeared in the 20th Century–Fox house organ, *Close-Ups*, entitled "Promote from the Ranks." Addressing the heads of various studio departments, the article asked, "When you are adding new members to your departments, why not give consideration to the young men who are working as office boys?" It continued:

> The studio office boys are selected with great care and an eye to their possible advancement into responsible positions. These boys, by the same token, accept office boy jobs in the hopes that their abilities will become recognized, and will entitle them to preferment in the matter of promotions. Some of them have been in the same jobs with Metro-Goldwyn-Mayer for several years. They have seen outsiders brought into better positions at which they deserve a chance. Why not give them a thought when an opening occurs, Mr. Department Head? Union regulations permit the various studio crafts to take on apprentices. The boy who delivers your mail, runs your errands, might make a good man for some technical department if given a chance at apprenticeship.[73]

The editorial's effusive concern over the plight of the office boy languishing in the same job for several year apparently doesn't spill over to women in nearby clerical positions, or to secretaries who often worked in closer proximity to studios' creative elite than office boys and carried out higher-level tasks in service of their employers. Today's girl Friday was more likely to be seen as tomorrow's wife and mother—an assumption that speaks to the extent to which larger cultural currents had insinuated themselves into studios' and workers' conceptions of women's work.

Work Wives: The Secretary as Constructed by Studio and Wider Culture

Once established, societal gender norms both in and outside of the workplace require constant reinforcement to remain in place. This reinforcement is delivered through the language with which genders are discussed, the way men and women are addressed, and recognizable signs and symbols of gender-based hierarchy. In studio-era Hollywood, studio culture provided continuous reminders of how female employees should look, act, and behave. This was particularly pronounced in the case of studio secretaries, with their high level of visibility among workers and frequent isolation from other female movie workers in the offices of movie makers. Using secretaries as primary examples, the remainder of this chapter unpacks the ways in which the culture that surrounded female movie workers shaped their professional identities, their behavior, and their prospects outside of feminized fields.

Secretaries found their way to the desks of employers through a variety of channels. Some were hired directly by a particular individual or department, often through referrals or acts of nepotism. Others answered job ads at studio personnel departments, or were placed through temp agencies.[74] Most would-be secretaries were expected to be proficient typists and high school graduates. To work for high-profile employers, they might face additional educational and training requirements, such as a year in a commercial course, a business degree, and several years of secretarial experience.[75] Though they were detailed in personnel files, these considerable duties for which secretaries were hired went largely undiscussed in studio culture. Secretaries were far more often represented in terms of gender and sexuality than in the context of their professional skills. Studio culture reinforced traditional views of women's natural sphere (the home) and her natural goals (marriage, children, and femininity). Moreover, where they concerned secretaries, studio-produced films themselves perpetuated these values.

Secretaries in Fiction

Hollywood films of the era typically defined secretary characters through their sexual difference from male employers. "Secretary" often signified sex—a female body in service of male needs. Writing of fiction, Leah Price and Pamela Thurschwell have proposed, "If the secretary's mind is sometimes pictured as disposable, a machine for mechanical reproduction, her body is simultaneously accessible on-site."[76] In the world of the film, secretaries are not only sexed, but also sexualized, and even oversexed. Virtuous typist and secretary characters were frequently offset by highly sexual, seductive, home-wrecking ones, like Barbara Stanwyck's "Lily" in *Baby Face* (1933), who literally sleeps her way to the top of a company, taking lovers floor by floor, until she lands in the lap of luxury. Though Lily makes a fairly straightforward trade of goods for services, because she uses her sexuality to her own advantage she is cast as a bad woman, particularly in the censored version of the film.

Less sexual but no less feminine was the secretarial archetype of the loyal, often long-suffering "girl Friday" or "office wife," a nurturing, feminine, workplace companion for lawyers, private eyes, and businessmen. Such characters—paragons of womanly workplace support who reinforced the studio's own "real-life" norms of virtue via self-sacrifice—can be found

Fig. 21. Production still from *The Maltese Falcon* (1941) depicting secretary Effie Perine (Lee Patrick) and employer Sam Spade (Humphrey Bogart). (Courtesy of the Academy of Motion Picture Arts and Sciences.)

in an array of 1930s and 1940s studio films, from Warner's *The Office Wife* (1930) to RKO's *Mr. Blandings Builds His Dream House* (1948). Studio films, as historian Susan Elizabeth Dalton explains, typically depicted the secretary as "the woman who did the most for a man," via a variety of roles: "She was a slave who did a man's work; a conscience who watched a man's soul; a mother who guarded a man's health; and, at the end of the last reel, a lover when she was 'discovered' by her boss."[77] One of the most popular iterations of the "office wife" story casts the plucky, truehearted secretary as the primary female love interest of an unattached male employer. Lynn Peril observes that in novels featuring secretary protagonists, the vast majority used the office as mere backdrop, against which "the heroine took some shorthand, typed a letter or two, then a married the boss—after many trials and tribulations, of course."[78] Even Silvia Schulman's *I Lost My Girlish Laughter*, the novel based on her experiences working for David O. Selznick and then leaving him to marry Ring Lardner Jr., uses the same Cinderella story structure. However, as its title implies, the book is hardly a ringing endorsement of the men studio secretaries served. Indeed, much of its story serves as an indictment of the protagonist's lascivious, vindictive boss—a character clearly patterned after Selznick—and his powerful friends.[79]

"Real" Secretaries (as Depicted by Newsletters)

Studio newsletters, which recorded much of movie workers' professional culture in the 1930s and '40s, were seldom work centered where female employees were concerned—and were almost never critical of the studio. Their descriptions of female workers—penned by the women themselves or by their male coworkers—attest to an almost singular focus on their femininity, glossing over their work duties in favor of detailing their smiles, looks, love interests, and propensities to giggle, chatter, and gossip. Reports on and from secretaries focused particularly on feminine appearance and behavior (for example, clothing and hairstyle—which girl had the best smile or laughed the most), or secretaries' romantic prospects. Many issues resemble a marriage mart more than the internal newsletter of a large company. Nearly every "Stenographic" or "Secretarial" column in the *Warner Club News*, for example, contains pictures and reports of recent weddings, and gossip about desired ones.[80] Although female studio employees were not prohibited from working after marriage and, indeed, many did, the goals of marriage and a family are repeatedly asserted by reports like the following from the MGM newsletter: "Virginia Elston and Joe Richardson, both former readers—plighted a troth and performed a marriage merger recently, just like in the story books. Joe is now a hardworking script clerk on the lot. Virginia is ready to answer the census taker with: Occupation, housewife."[81]

Similar announcements, as well as speculation about which secretary had a beau and who was ripe for one, are made in numerous RKO and MGM newsletters, with relationships between secretaries and the men to whom they are assigned framed with similar language.[82] In one 1939 issue, RKO's "Notes from Stenographic" consisted of stories of weddings, engagements, and "happy recent assignments" of secretaries to new bosses, all listed together. At MGM, secretaries were polled about their bosses ("Who knows bosses like secretaries know bosses?"); the results concern the sorts of physical or emotional characteristics that a wife might value in a husband (handsomest, wittiest, best dressed, most glamorous, most cheerful). The award for most absent-minded boss was a six-way tie in which one secretary reportedly voted for "Everybody's boss." The cheerful commiseration of these "work wives" regarding their men both legitimizes and necessitates women's prescribed domestic role as the detail-oriented helper to her (office) family's chief executive.

Commensurate with these social concerns, newsletters described women's appearances more often than men's—even in stories of their workplace accomplishments. One dispatch announces the sale of a script for production at another studio by "Jean Baker, attractive, elegant secretary to Art

Fig. 22. Typical "Stenographic" column from *Warner Club News*, May 1944. (Courtesy of the Academy of Motion Picture Arts and Sciences.)

Director Johnny Hughes."[83] Other columns solicited wives for unmarried male workers, such as Vic Raven, a painter whose "specifications for a wife calls for a young, good-looking blonde, brunette or red hair, good cook and excellent housekeeper. Girls, the line forms on the right. Vic will receive applications daily at noon in the paint shop. Pls Apply in person."[84]

Secretaries and other female workers filled the gamut of women's roles, from mother and companion to morale booster and sex object. During World War II, much like female stars who were expected to pose as pin-ups in studio publicity photos, women workers' made their contributions to the war effort mostly through performed femininity and sexuality, posing for their own pin-up photos in studio newsletters, entertaining soldiers at the Hollywood canteen, and adopting enlisted studio workers as pen pals ("Now, girls of MGM, we can do our share. . . . Yes let's each adopt one of our own draftees in training").[85] But the female worker was not the pin-up girl only when her country was at war. Women were frequently posted in or around male-dominated departments as clerical workers or in nearby women's labor sectors (for example, the film lab, the ink and paint department), and men and women socialized in shared studio spaces such as commissaries and cafés. And per studios' esprit de corps mythology and wider cultural norms of the day, women were expected to be willing recipients of their male coworkers' gazes.

Along similar lines, the secretarial and other columns continually discussed women's looks and dress. One dispatch from the timekeepers at Warner Bros. describes the department's male workers all a dither that one of its female clerical workers, Gertrude Archier, wore red slacks when called in to work on a weekend. The report ends, "Gertrude is now the most popular girl in the time office and the boys hope they have to work every Sunday."[86] The generally high interest in female movie workers' physical characteristics and attire suggests that male colleagues viewed them not only as future wives and mothers but also as fetishized objects of distraction, amusement, and sexual desire.

Whether conscious or unconscious, this focus on these trappings of femininity probably served a larger, psychological purpose related to company morale. Women's presence in the studio workplace inherently challenged masculine authority, triggering what might be understood as a reaction-formation—a defensive process that eliminates or offsets an anxiety-producing threat by exaggerating its opposite—in this case, amplifying gender differences.[87] Hyperbolizing women's sexuality and femininity neutralized the implicit threat they posed. Hyperbolic femininity was constructed from without—the way women were hailed, discussed, described by male colleagues—but also from within, by the workers themselves. Women freely participated in this culture of surface distractions, frequently reporting on coworkers' looks or an "attractive new hairdress," placing want ads in the secretaries column for "handsome men, must own tuxedo, to escort girls" to the studios' formal affairs.[88] In so doing, they guarded against the loss of their femininity, still considered an ever-looming danger of working in a man's world, and thus helped male colleagues maintain their sense of the workplace as theirs. Gender and sexuality were so central to women's self-conception as studio workers that they served as packaging for everything else. The newsletters' hundreds—perhaps thousands—of references to the physical beauty or feminine qualities of this or that secretary attest to the reality that identity as a woman was almost always privileged over workplace identity as a (productive) laborer.

The same hierarchy of values was evident in studios' promotions of high-profile female workers such as Virginia Van Upp, who, when she ascended to executive producer at Columbia, was rephotographed by its publicity department. As Lizzie Francke explains, the department withdrew all photographs previously used in promotion—which showed a bespectacled Van Upp posed near a film camera—replacing them with a newer photograph of Van Upp, sans glasses and posed with a flower tucked behind one ear as if to underline that, although powerful, she was "unthreateningly feminine."[89] On a more overtly critical note, Joan Harrison, another powerful woman at studios in the 1940s, wrote an essay for the *Hollywood Reporter* entitled "Why I Envy Male Producers," which asserted that men spent more time engaged in work and leisure pursuits because they didn't have to worry about their appearance, while she, as a female producer, had to spend hours at the hairdresser and was judged for shadows under her eyes rather than congratulated on them as a sign of hard work.[90] Other female workers' private reflections on studio culture evince a similar negotiation between their individual identities and workplace expectations related to their gender.

Secretaries as Symbols

The performance of femininity served a symbolic purpose in the large, self-styled mini-cities that studios had become by the late 1920s. As chapter 2 explained, feminine assets—from the machinelike bodies of seamstresses to the sexualized bodies of chorus girls—adorned the studio promotional materials to substantiate the claim that they were full-service, self-contained movie factories. Though they lacked positions of authority themselves, studio secretaries—as presented in visual material—served as potent symbols of the studio's power and the commanding positions of the individuals they served. Much like secretaries depicted in popular advertising of the time, studio secretaries were framed as "both the object and the basis of men's power and control," existing "to operate men's machines and to service men—in ways that are, by implication, rather intimate."[91]

Secretaries were visually and spatially central to the office's mise-en-scène, situated as part of the executive's art of intimidation. In the recollections of employees who were kept waiting by executives as a demonstration of status, part of the ritual's humiliation was that it took place under the watchful eyes of the executives' secretaries.[92] And, indeed, once a secretary came to signify her employer—as so many did in service of executives for whom appearances were often as important as skill—it was hard to reverse the process. For example, despite her thirty-year career as a writer and producer, Joan Harrison is invariably described by colleagues as "Hitchcock's former secretary" in their accounts to historians. Though the director himself took pains to elevate her from that status, that effort didn't sit well with men like Charles Bennett, who insisted, years after he and Harrison shared screenwriting credit on *Foreign Correspondent* (1940), that she was just "our secretary" and had never come up with a single idea on the film.[93]

With secretaries and other women workers on hand to fill out the studio city's simulation of the outside world, studio lots were discursively framed as homes away from home, complete with mothers, wives, and figures of sexual fulfillment. By "owning" these sources of femininity, studios built in some of the favorable working conditions promised by the often-farfetched studio-as-city promotion. The implication: studio workers (male) could spend long days and nights in production without worrying about how they would be fed, nursed, nurtured, and so forth. Like the vast array of guns that studio armories acquired on the chance they might be needed for a future film, these requisite feminine assets were already "in stock."

"Anything That Had a Skirt Had Better Be Careful": Studio Work Culture's Cost to Female Employees

Women studio-wide were expected to comply with universal gender norms dictating traditional feminine appearance (for example, dress and hairstyle) and behavior (for example, subservience, outward repudiation of "masculine" traits like ambition or competitiveness). At Disney, female animation workers were required to wear skirts and dresses. In 1958, a painter who showed up in a Katharine Hepburn–inspired pantsuit was fired.[94] Dress codes, official or not, were often directly communicated to secretaries; work for high-status movie makers meant dressing in accordance with their station. When Samuel Goldwyn's general manager hired Valeria Beletti to work as the mogul's secretary, he told her that she "would have to look very smart and dress well" and loaned her money against later paychecks for the purchase of suitable wardrobe. Beletti wrote of the experience, "I feel as if I'm a different person entirely. For once in my life I bought real stylish clothes and they do make a difference. Of course, I have to keep my hair marcelled, but in view of the salary I am being paid, I can easily do it."[95] But Beletti soon tired of this required attention to appearance, echoing Harrison's envy of the male producer when she wrote, "That's one draw back about this job. I have to look nice, and that's so hard for me because I hate to shop and worry about clothes." Producers' secretaries at MGM earned a high salary for the field (up to $65 for a forty-hour week plus overtime), in part because they were expected to dress nicely. Emily Torchia recalled, "We didn't wear slacks; if we had, we'd have been sent home."[96] During her tenure as an MGM secretary in the late 1930s, Torchia wrote a fashion column called "Style Scoops." At RKO costume designer Renié penned a similar column, "Fashion's Fancy," advising her readers in February 1941, "You are to look as feminine as possible and the new hats for this purpose are lovely and flattering."[97]

Feminine behavior was guided only slightly less directly than feminine dress. Harry Cohn attributed Virginia Van Upp's success in his organization to her ability to "survive in a man's world without losing her femininity."[98] Transgressors of feminine behavior norms received less praise and more censure, whether professional (in the form of promotions not received) or social. Though Samuel Marx admired Kate Corbaley (his assistant and the de facto head of his story department), he "had no doubt she was strongly feminist," and observed in light of this fact that "few of the males on the writing staff were her friends."[99] Instead, she shared a close friendship with screenwriter Frances Marion, largely behind the door of her office, which would be "closed to the world so they might enjoy their 'private gossip,' as Kate described those sessions."[100]

Studios sanctioned—and even encouraged—certain types of gossip as an acceptable feminine indulgence. Importantly, however, while some gossip was idle distraction, other chatter operated as a mode of social containment to punish transgressive female behavior. Secretaries at studios were reputed gossips, calling each other to "alleviate numbing boredom" by relating "fact, half-truth, and falsehood."[101] Meta Carpenter was on the receiving end of their telephone game after the departure from 20th Century–Fox of the well-known writer with whom she had begun a romance; in her words, "At least twenty secretaries now in the commissary would know that William Faulkner had left for Mississippi the day before, ending his affair with the Southern girl who worked for Howard Hawks; by clock-out time, half a hundred would have heard it." She later described the sense of isolation she felt in the norm-enforcing studio rumor mill: "Braving the crowded Café de Paris for lunch—Was anyone staring at me?—Those two writers with their heads together there, did they recognize me as the blonde who was always with William Faulkner at Musso & Frank's? . . . I darted my eyes right and left and saw, or imagined it—a number of heads turned my way."[102] Male employees also spread studio gossip. But female movie workers were simultaneously blamed for gossip and encouraged to indulge in it; as such, they reinforced cultural notions of women's behavior while undermining their potential to be promoted out of feminized sectors: if gossip—according to the culture's narrative—was an inborn feminine weakness, female movie workers were fundamentally unfit for dignified business roles.

A notable exception to this rule could be found in the para-industry of professional gossip and tabloid journalism.[103] Some rumors—related to stars and other high-status employees—served studio interests, functioning as a loosely managed form of viral marketing. As Carpenter explains, "There was an incredible network of rumor and exaggeration from which professional radio gossipmongers mined most of the inside news that entranced the public."[104] May Wale Brown echoed this sentiment, saying, "When working on movie sets, gossip swirled around me in a continuous flow," and the fear that so often fueled it, in turn, fed the promotional machine.[105] The inextricable connection between women and gossip not only legitimized but also elevated the work of female columnists like Louella Parsons and Hedda Hopper, whose femininity was framed as an additional qualification for the work of peddling rumors. As actress Evelyn Keyes explained, Parsons and Hopper were accepted in their roles because "women were supposed to be catty to each other. . . . It was our programming, like pink for girls. Gossip columnists were bitches—everybody knew that."[106] Through their feminine performance as gossipmongers, Parsons and Hopper ascended to levels of great power within the industry, fulfilling what studios viewed as an essential promotional function.

Studio culture also by turns exploited and regulated female sexuality. It was the female movie worker's job to respond to unwanted male attention without offending. Of course, this was the case outside of studios as well, as reflected by numerous secretarial manuals, which—as early as 1916—advised that the average woman was "quite capable of freezing an undesired admirer into a state of respectful good sense, without even losing her job in the process," and that in "a case where she can not get rid of the attentions, she can, as a last resort get rid of the job. . . . She need never be kissed twice against her will."[107] Manuals from the 1930s and 1940s counseled secretaries to put off married men by bringing up subjects that would remind them of domesticity, to "pay no attention to personal remarks, pats or other approaches. Pass them off and keep quiet," or to accept a date but repel the suitor by dressing frumpishly.[108] As late as the 1970s, secretaries were advised to give a boss who groped them the brush-off in a way that wouldn't embarrass him, or to administer a slap if the situation demanded—but only in private.[109]

But perhaps even more so than in traditional 9-to-5 offices, at studios, the onus for neutralizing expressions of masculine desire—from harmless to downright harassing—was on its female object. By studio secretaries' own accounts, sexual advances were expected to come with the territory, to be endured to the best of one's ability. Early in her stint at Goldwyn, Valeria Beletti reported cheerfully to a friend back home, "No one but the art director, Mr. Anton Grot, has gotten too friendly as yet. Mr. Grot tried once to kiss me but my guess is that he won't try again."[110] Grot continued to make advances, but she was able to make him "behave" by keeping her distance. Within months, she had a handbook-worthy experience involving Assistant Director Jimmy Dugan, in which a simple "no" was not accepted, even when pleasingly delivered:

> I know he's married and of course I don't want to be involved in any affair with a married man. He tried to tell me there wouldn't be any harm in his taking me out occasionally but I sternly refused to listen to him. I told him that with all the extra girls running after him ready to give themselves up to him for the sake of a little part in some picture, why should he want to take me out when he knew I wasn't a good sport. This is what he said: "Listen here. I've had enough of them—the very sight of girls of that sort repulses me. If I take a girl out I want one that I know I can talk to and don't have to make love to. I couldn't do that with those girls, but I can with you, because I know you're good and haven't been out with a lot of men."[111]

The no-win nature of Beletti's situation is clear: her chasteness stokes more pursuit; yet the consequences of giving into that pursuit (that is, being a "good sport") are that she will eventually be found "repulsive."

Though most secretaries withstood the attentions of male coworkers as part of the job, few relished the experience. Marcella Rabwin's memoir contains little criticism of her former employer, David O. Selznick, but grants, "The sexual use of candidates for studio jobs was almost universally practiced in the industry, and David was no exception. . . . I was tolerant of his misbehavior; my prior experience had prepared me."[112] Indeed, Rabwin had met with her own share of male attention—some consensual, but far more of it unwanted and unsettling. According to Rabwin, when Darryl Zanuck spotted her working for writer Arthur Caesar, he asked her to move to his office "as assistant to his private, private secretary," whose primary function, Rabwin knew, was recording "information about the choicest girls on the lot." Attempts to beg off were fruitless: when she told him she was happy where she was, she recalls, "He fired Caesar and gave me two options: accepting or leaving. I should have left, but I hadn't yet found the courage to be out of work. Within two months, his chases around his desk, polo stick ever in hand, became so frightening that one night I ran out of his office into the hall screaming, and never returned."[113] Rabwin related this story fifty years after it happened, which may explain its close resemblance to a woman-in-jeopardy movie. However, since the experience seemed to have disturbed the author greatly, one can interpret her memory as accurate at least in the sense it provides of how it *felt* to work for Zanuck for those months. And this is hardly the only instance of such harassment related by secretaries and other female workers in memoirs and other accounts of their work. Indeed, the Zanuck incident is not even *Rabwin's* only experience of harassing behavior on the job. Rabwin found work in Myron Selznick's organization, but in the office of "the wrong member of his firm"—executive George Volck—whom she described as "venomous, Teutonic, and cruel," and who attempted to lure Rabwin into homoerotic encounters with another secretary. When Rabwin refused, he began a campaign of verbal abuse that ended in her resignation.[114]

Meta Carpenter enjoyed harmless mutual flirtations with Maurice Chevalier while on location with employer Howard Hawks but looked less kindly on the advances of writer Gene Fowler, who appeared at her hotel room under the pretense of helping transcribe dictation he'd given her earlier, then "ordered a dozen martinis brought up, drank them all himself when I refused to join him, and at the end of the afternoon became abusive when I showed no interest in letting him lead me into the bedroom."[115]

Such behavior could be found not only in offices and hotel rooms but also on film sets. Alma Young worked as a script clerk for director Henry Lehrman, whom she remembered as being nicknamed "Suicide Lehrman" because "he ruined so many people, and so many people committed suicide. And anything that had a skirt had better be careful." She recalled receiving the assignment from production manager Bill Koenig, who was concerned for her safety around the director: "He said, 'Now this guy is murder. The property man will pick you up in the morning. He will have lunch with you. He will bring you back at night. And when you have to go to the restroom, he will go with you.' I said 'O.K.' What else can you say?"[116] Such stories elucidate the precariousness of women's positions. Though most secretaries reported a similar mix of consensual and nonconsensual, gentlemanly and ungentlemanly treatment from superiors, the unifying characteristic of all of their accounts was lack of control over their own working conditions. If a male coworker interceded on the secretary's behalf, it was purely by his own choice and at her peril more than his, as she would bear the consequences of whatever resulted—whether through

loss of her virtue or of her job. Sexual exploitation was especially danger-
ous in the small, self-contained cities that studios had become. Isolated
from the outside world, with their own rules, police forces, and authority
(the very producers and moguls who abused workers in this way), studios
were small fiefdoms, where power and perks permitted cherished producers and executives to operate free from oversight.

Unusual business hours and locations further blurred the line between
secretary and wife, chorus girl and girlfriend, distorting the boundary
between work and leisure time. Copious alcohol added to the confusion.
Valeria Beletti felt pressured to drink on the job in order to win her male
colleagues' cooperation during regular hours, saying, "What I'd like to
know is how I'm expected to be able to work and drink at the same time. . . .
You know one must be a little sociable with their co-workers—if I didn't,
I'd get a pretty bad reputation and I wouldn't be able to get anyone to do
anything for me."[117]

At times, more was demanded of women than social or professional
appeasement. Indeed, the permissive environment and the excesses of
male authority could lead to sexual violence and abuse. Samuel Marx wit-
nessed "X-rated" behavior at MGM Christmas parties in the 1930s, which
were stocked with booze to help workers "blow off steam."[118] Unfortunately
blowing off steam was not a voluntary activity for all MGM employees.
Patricia Douglas, a twenty-year-old dancer, was hired for what she thought
was work on a movie, only to discover, after being costumed, made up, and
transported with 119 other girls to the Roach Ranch, she was to play hostess
to the studio's regional salesmen at their 1937 convention.[119] The banquet
hall featured abundant scotch and five hundred cases of champagne for the
three hundred men in attendance. Things quickly got out of hand as the
drunken conventioneers took advantage of the stocked pond of women
who had seemingly been provided for them by studio boss Louis B. Mayer.
A few days earlier, Mayer had advertised this perk in receiving the conventioneers; flanked by starlets, he stated, "These lovely girls—and you
have the finest of them—greet you. . . . And that's to show you how we feel
about you, and the kind of a good time that's ahead of you. . . . Anything
you want."[120] Waiter Oscar Buddlin later testified to seeing girls "get up and
move from the tables because the men were attempting to molest them."
Another waiter stated, "The party was the worst, the wildest, and the rot-
tenest I have ever seen," and that "the men's attitude was very rough. They
were running their hands over the girls' bodies, and tried to force liquor on
the girls."[121] A conventioneer named David Ross pestered Douglas all night
(in the bathroom, she remarked to an attendant, "I've got a man and he's
really sticking"). Ross took offense at her repeated protestations as a teeto-
taler and held her down so that his friends could make her drink. Sixty-five

years later, Douglas recounted her experiences that night: "One pinched my nose so I'd have to open my mouth to breathe. Then they poured a whole glassful of scotch *and* champagne down my throat. Oh, I fought! But they thought it was funny. I remember a lot of laughter." Douglas fled to the bathroom to vomit, but when she stepped outside for some air, Ross grabbed her, held her down, and raped her, saying "Make a sound and you'll never breathe again."[122]

Badly beaten and in shock, Douglas was sent home after seeing an on-call doctor affiliated with the studio. She later told an MGM cashier what had happened, but was offered only her day's pay: $7.50. When she brought suit against Ross, Hal Roach, and MGM, the studio responded by hiring Pinkerton detectives to canvass her acquaintances to obtain proof that she wasn't a virgin, as she'd claimed. According to Budd Schulberg, at that time, MGM "owned *everyone*—the D.A., the L.A.P.D. They *ran* this place."[123] Predictably, the lawsuit was dismissed. Parking attendant Clayton Soth, who had identified David Ross as Douglas's attacker, changed his story on the stand and was, per Stenn, given a lifetime job driving cars at the studio. In court, Mendel Silberberg, Mayer's personal attorney, came to Ross and MGM's defense. Indicating the plaintiff, he said, "*Look* at her. Who would want *her*."[124] The studio generally went unnamed in the papers, which printed not only Douglas's picture, but also her name and address.[125] Douglas, who claimed the party and the events that followed ruined her life, insisted that she had never wanted money from her suit, stating, "I just wanted to make them stop having those parties."[126]

Though many who worked at the studio at this time privately expressed disgust at such incidents, this behavior was hardly discouraged by MGM's management, which "kept a group of girls on hand for the pleasure of visiting dignitaries"—not to mention executives whose behavior was often anything but model.[127] Eddie Mannix's repeated brutal acts of physical violence sent his wife and at least one of his mistresses to the hospital.[128] L. B. Mayer himself indulged in affairs with a series of chorus girls and starlets, whom Frances Marion referred to as "silver platter girls" (for how they delivered themselves—metaphorically, to Mayer) and "moos" (because, on delivery, they became his "sacred cows," whose names he suggested in casting).[129] Clearly, studio management not only tolerated the actively sexualized environment but encouraged it.

Through actions and words, executives modeled the very behavior that produced and perpetuated this climate, effectively legitimizing it as part of studios' institutional structures. Not surprisingly, then, there was pervasive tolerance for "boys being boys" at MGM and other studios. When word got around about cattle calls for musicals, "male workers found an excuse to be in that vicinity."[130] Said Dmytryk of this practice at Universal, "The chorus girls obviously had to dress in tight shorts to show their figure, and men from all over the studio would be standing there leering."[131] For Peggy Montgomery, an extra who began work at MGM at the age of sixteen, this climate meant

> a constant air of being pursued. All the men tended to try to break women down. These were very aggressive men. Twice, I was asked to go to be interviewed and the guy got up and said "Well, let's see your legs." And you pull up your skirt and he said "turn around, sweetie, pull 'em up higher." And then he'd say, "let's see how you feel." And then he'd walk around the desk and grab you. You couldn't go to the citizen news and say "You know, Mr. So-and-So did this to me at MGM." No way . . . I mean this is no exaggeration it was one of the laws I learned very early on.[132]

These recollections cast the discursively framed "studio-as-city" in a disturbing light. With not only their own nurses and teachers, but also their own doctors and policemen, enclosed studio cities didn't just keep outsiders away—they held employees captive inside. There, secretaries were expected to be loyal to employers even if doing so operated against their own interests, all in an actively hostile, often aggressively sexualized environment that hyperbolized their gender difference. Many worked at the same studios for years as their bosses' cheerleaders and staunchest supporters, their professional fortunes rising and falling with those of their employers and having little chance of promotion, especially as an executive or to produce or direct.

Conclusion

Women continued to carry out the bulk of the clerical labor in Hollywood long after the end of the studio era. In 1980, the *Los Angeles Times* reported that Karen Neumeyer, a newly appointed union leader for Local 174, the Office and Professional Employees International Union that, by then, represented 2,300 "clerks, data processors, programmers and secretaries at MGM, Columbia, Universal, 20th Century–Fox and a number of film labs." The article's author states that "if female directors are a rare breed," female union leaders are "an unknown species" in the movie industry, "a business dominated by guilds and unions at almost every level." Yet, he explains, "more than 75% of its workers are women, so, *in a sense*, it's appropriate that its leader, the person responsible for negotiating contracts and handling grievances, be a woman" [emphasis mine].[133] The reporter goes on to say that Neumeyer is "a soft-spoken attorney in her 30s whose manner is as unassuming as her understated business office" and who once worked as a secretary. When she is finally quoted directly in the lower third of the article, Neumeyer dismisses the question of whether it is appropriate for a woman to head the union ("I think that's up to the electorate") and instead explains that "secretaries need a union for the same reason that other people need a union, because unions increase the economic benefits of their members," and adding, "Women in the U.S. today make 59 cents for every dollar that a man makes. The non-union woman makes almost a hundred dollars less per week than the unionized woman. We need unions in this industry for secretaries."

It's not surprising that a reporter in 1980 buried this lede—that secretaries in specific and women in general need unions—under five paragraphs devoted to the mystery of how a woman wound up in a leadership role and to authenticating her soft-spoken femininity (as if to reassure the reader that her leadership role hadn't made her turn "hard"). Even today, stories ostensibly about women's professional success are framed as stories about their gender, or with equal emphasis on their accomplishments as wives, mothers, and so on, as if to balance out their workplace achievements. The 2013 *New York Times* obituary for Yvonne Brill, who invented a propulsion system to keep satellites in their orbits, originally began "She made a mean beef stroganoff, followed her husband from job to job and took eight years off from work to raise three children," going on to explain that Brill's son thought she was the "world's best mom," before a second paragraph revealed that Brill also happened to be a brilliant rocket scientist.[134]

Neumeyer and Brill succeeded in male-dominated fields, so perhaps the authors' focus on their gender is understandable. Yet it is nonetheless significant that these authors feel compelled to discuss their female subjects' gender before they can consider their work. In accounts of women in the classical Hollywood era, the fact of their gender doesn't just come first, but seems to eclipse any individual skills, talents, characteristics, or achievements to such a degree that they're barely visible. Until very recently, women could participate in the professional sphere only if they did so within a frame of femininity, whether they actively shaped that frame themselves, say, by posing as pin-up girls, or merely acquiesced to the frame their culture imposed on them by complying with unwritten rules about skirts and smiling. Gender and femininity framed not only women's workplace contributions in feminized labor sectors at the studios, but also their advancement beyond them, because in the sex-segregated film industry, it was from the women's sectors described throughout the previous two chapters that women first entered film production fields that were not strictly for women. As the next chapter explains, in spite of tight restrictions on their professional mobility and often troubling working conditions, some women still managed to operate sub rosa in a creative capacity from feminized sectors. Doing so eventually helped some of them succeed out of strictly feminized roles. However, their advancement and their success in new fields would require them to negotiate the same workplace culture and face the same gender-based expectations, with consequences to their pay, status, and professional identities.

Fig. 23. Peggy Robertson and Alfred Hitchcock on the set of *Vertigo* (1958). (Courtesy of the Academy of Motion Picture Arts and Sciences.)

4

"His Acolyte on the Altar of Cinema"

The Studio Secretary's Creative Service

His wife left town and he said to me "I need you to go home to my house every night after the nanny leaves and put my kids to bed." And there wasn't really, like, I wasn't given an option.

—Anonymous TV writer's assistant, 2007

A 1915 issue of the *Moving Picture World* reported the hiring of Mrs. Robert T. Haines as "social mentor" of the Equitable Motion Picture Studio to "sponsor for the social correctness" of films, advise on costumes, settings, social etiquette, and other "little artistic niceties which are so often overlooked by men in their less careful regard for detail." Haines told the reporter that a director might save himself much anguish if he "could rely on the judgment of a sane, well-balanced woman with artistic sense and clarity of mind to assist him." She did not believe many women could direct but explained that, in turn, "neither is a man able to feel the little things, nor does he possess the intuition that woman does." However, given the number of so-called feminine strengths cited as necessary to keeping a film on track (judgment, artistic sense and clarity, balance, intuition, eye for such "details" as costumes, sets, acting), the article makes a stronger case for female directors than against them.¹ Despite purported concern with "little things," the social mentor's role, as described, was so broad in scope—taking in the whole process of production—that out of context it might easily be mistaken for that of a director. Haines was not the first woman to steer the ship of production under the legitimizing umbrella of feminine intuition, social sensibility, and attention to detail, nor would she be the last. However, as women's prospects in production narrowed according to gender norms, women were increasingly expected to use such skills to advance male employers' creative projects rather than their own.

As earlier chapters explained, studios' adoption of scientific management principles and other established practices from major industries shifted clerical work in importance from the margins to the center of film production. At the same time, female clerical workers, the studios' primary paper workers, were marginalized in their potential to succeed beyond the typewriter. For them, promotion was a rare to nonexistent prospect, often limited to jobs in planning departments, and dependent on the their willingness to play their feminine part and add value to their labor through qualities and skills that supposedly came naturally to them under gender-normative cultural assumptions, all in a work culture that hyperbolized their femininity to the degree that sexual harassment and assault were practically a fait accompli. Gender-related expectations didn't simply exist around the edges of women's work, say, on their lunch breaks or at the office Christmas party. Whether it was directly stated or not, playing this feminine role was part and parcel of the job itself and refusal to do so was at very least a severe handicap on women's professional success and potential to advance into the narrow range of fields that were still open to them, and at most grounds for dismissal. This was especially true for women in feminized labor sectors like the secretarial pool, who were viewed as replaceable and whose working conditions depended entirely on their mostly male employers.

For these reasons, women's work at studios must be defined as a combination of *both* the explicit labor they were assigned on the basis of gender—typing, sewing, inking and painting—*and* the implicit "shadow" labor—the interpersonal competencies, gender performativity, and emotion work their jobs required.² This chapter serves to both elucidate and assign credit for these implicit aspects of women's work, referred to collectively in this chapter as "feminine roles," as uncredited, unpaid labor. Recognizing feminine roles as compulsory components of women's work is one way to ensure that women won't feel obliged to continue playing them, as they often still do in order to make their presence more palatable in male-dominated production sectors (for example, cinematography, comedy writing). In the chapter's case studies of five notable secretaries and assistants, constructed through their accounts of the work in combination with existing histories, the concept of creative service—fashioned for the purpose from relevant gender and labor studies frameworks—describes the basic, shared function of secretarial labor and, to some degree, all women's labor at studios, serves as a spectrum along which to align the different types of work women did as secretaries to movie makers, and illustrates how other feminized film work may be similarly linked. The case studies also illustrate the high level of creative and managerial agency some women achieved despite their industry's structural inequalities.

Women's Secretarial Work as Creative Service

A Warner Bros. personnel form from the 1940s described the executive secretary's duties as follows:

> Takes dictation in shorthand of confidential and general correspondence, including data of a legal nature. Transcribes dictated material, types miscellaneous reports and other data. Maintains a weekly record of stories and scripts. Prepares and types a weekly script progress report. Maintains files. Answers telephone calls and gives requested information when feasible. Relieves the Executive of minor administrative duties by taking care of details that do not require his personal attention.[3]

Yet even as the form inventories the job's visible clerical tasks, it leaves many obligations unnamed. Though secretarial work technically consisted largely of clerical and administrative tasks, mastery of these skills was by no means a guarantee of success in the field.

Secretarial handbooks and training manuals, though not specific to the film industry, gave more complete accountings of general requirements at the time. The following passage from a 1926 vocational manual cites clerical requirements in its description of stenographic work, distinguishing secretarial work from stenography by describing the former's largely extra-clerical requirements:

> She stands as a buffer between him and as many troubles as she can. She shoos away the unwanted caller as politely as possible. She finds the papers her employer has himself mislaid, and presents them to him without an "I told you so." . . . She sometimes is forced to listen to discussions of his intimate affairs, but must be able to maintain an impersonal attitude. Because of the very many hours, the long periods of close consultation she must spend with her employer, she must

be careful never to arouse resentments on the part of his wife or gossip by fellow workers, for nothing is more quickly a certain road to failure.[4]

Though some of the work described is purely clerical, the references to communication, support, and the managing of the secretary's status vis-à-vis her employer indicate the field's underpinnings as a service profession.

Service has long been considered a defining characteristic of many female-dominated fields, yet the specific competencies required for this crucial job component are rarely formally established. In formal descriptions of the work, aptitudes like those described in the previous passage— dealing with an employer's moods, presenting mislaid papers without judgment—are often referenced obliquely as "soft" or "interpersonal" skills and thus exempted from definition.

The sociologist Arlie Hochschild uses the term "emotional labor" to objectify those parts of service jobs in which workers are required to coordinate their feelings with their labor or in which "the emotional style of offering the service is part of the service itself."[5] Emotional labor jobs are those that entail face-to-face or voice-to-voice interaction, require workers to produce an emotional state in another person, and allow employers to exercise control over the emotional activities of employees.[6] For example, flight attendants in Hochschild's study reported that their airlines required them to smile and present cheerful facades to passengers regardless of their working conditions, how they personally felt, or how passengers treated them. At times, they actively suppressed their own emotions to provide this company-mandated atmosphere of cheer. Though both men and women do emotional labor, Hochschild explains that the two genders experience it in different ways. Men who specialize in emotional labor tend to specialize in the "bill collection side of it," wielding anger and making threats on behalf of the company, while women, and particularly the middle-class women who tend to work in service professions, specialize in "the flight attendant side," and are thought to manage expression and feeling not only better but more often than men do.[7] Furthermore, emotion work is more important or prominent for women in the professional sphere because, traditionally, they occupy a subordinate social stratum with less access to money, authority, or status in society. Women have commonly traded emotional labor, particularly that which "affirms, enhances, and celebrates the well-being and status of others," for economic support, often by "creating the emotional tone of social encounters: expressing joy at the Christmas presents others open, creating the sense of surprise at birthdays, or displaying alarm at the mouse in the kitchen."[8] Not only that, but women frequently "react to subordination by making defensive use of sexual beauty, charm, and relational skills," and these capacities are therefore more likely to be exploited commercially.[9] Thus, although they may do emotional labor for wages, men are more likely to be seen as possessing those skills by chance or as products of their individual talents, while female workers' emotional labor is seen as an essential quality of their gender, which specializes in "the 'womanly' art of living up to *private* emotional conventions."[10]

Like the flight attendants Hochschild studied, secretaries were expected to coordinate their feelings with their labor in ways that benefitted their employers or organization. Such enforced attitudes were not recognized as labor but were instead considered innate traits of women, who were naturally agreeable, nurturing, friendly, emotionally giving, et cetera, because of their roles as caregivers in the domestic sphere. One columnist in the 1940s explained that women made better secretaries because male bosses needed "steady respect, even admiration from those about him—and these things a woman, because she is a woman, can give him. . . . Women's success as a secretary was not due to masculine qualities in her, but the converse."[11] Another writer held that nurturing a male employer, in fact, helped a woman express her femininity, because she could put to use "'all the skills that most women naturally possess—language faculty, organizational ability, willingness to pay attention to fine detail, intuitiveness, tact, warmth in handling people."[12] Of course, these discussions barely mention the work for which secretaries were technically being paid. Though clerical duties were no doubt also important, based on accounts by successful studio secretaries, a new hire would be unlikely to reach the highest degree of success in her role unless she also met the job's implicit requirement that she perform as a woman.

Mervyn LeRoy devotes an entire chapter of *It Takes More Than Talent* to studio secretaries, and his description of what a good secretary does so perfectly illustrates the service and emotional labor aspects of work for movie makers that it warrants quoting at length:

> The first time I met [secretary Rose de Luca], I knew she had infinite tact—a way of making people feel good, a way of telling unpleasant facts pleasantly. In Hollywood, we are put in the position of saying no quite often, but there are ways of saying no. Rose has the gift of taking the sting out of refusals. On the other hand, she has a definitely firmness when dealing with phonies. . . . I don't care how fast Rose's shorthand is. . . . Her real value is that she can make friends for me, or enemies, just by the ways she handles a phone call, a letter, a visitor to my office. . . . The great difference between just any secretary and an outstanding one is the willingness to put more into her job than she gets out of it. A studio can't pay for what a good secretary does.

Money can't buy loyalty, good judgment, or native intelligence or
the selfless desire to work hard, to stay until the job is done, and
done well.[13]

As LeRoy defines it, the job is composed almost entirely of emotional
labor, which the truly great secretary offers without the expectation of
being appropriately compensated in return. In this passage, he seems to
have discovered the only thing in the world that studios—the proud pur-
chasers of armories, lumber mills, menageries, luxury cars, and movie
stars—can't buy. What the director is saying, however, is not that studios
can't pay secretaries more, but that the secretary's self-sacrifice wouldn't
mean as much if she did it for the money. Meanwhile, the secretary's benefit to the studio's bottom line is manifest in his anecdote about a message
from a designer to an art director that, because it wasn't tactfully delivered, took an hour-long phone call to soothe hurt feelings. LeRoy explains,
"This soothing cost the studio about five hundred dollars in time spent by
two high-salaried men," when, "if the message had been tactfully delivered
by a secretary adept at handling easily offended egos, the studio would
have saved money."[14] As if the job isn't already marked clearly enough as
women's work, LeRoy adds the additional qualifications that successful
applicants will be "pleasant-looking" and between the ages of twenty-three
and twenty-eight. Presumably, being young, pretty, and female helps the
secretary "make friends" and "take the sting out of refusals" for her boss.[15]
If these requirements seem a bit shocking, it's only because LeRoy has
written them down for others to read, not because they weren't typical of
the expectations women faced when they stepped into support roles.

Until it was pointed out that Ginger Rogers did everything Fred Astaire
did—only "backwards and in high heels"—the added skill her role required
was often overlooked precisely because Rogers played that role so well.[16]
Success depended on her ability to disguise the intense effort that went
into her dancing. Likewise, for many women in the studio era, successful
job performance depended on gracefully and covertly fulfilling one's role.
Film history offers its own version of this contradiction, not only confining
workers like the studio secretary almost entirely to anecdote and footnote,
but often describing important female figures in film as having succeeded
in spite of their gender. In fact, many women succeeded in affecting film
history *because* of gender-specific performance *as well as* performance in
all of the same areas as their male peers. To acknowledge this condition
of so many women's participation in film history does not diminish the
accomplishments of those who resisted and negotiated prescribed gender
roles. It simply gives credit where credit is due for what playing the feminine role at studios really was: *work.*

Inconspicuousness and erasure characterized much of the film-specific labor assigned to women at studios. Though many technical and craft workers were expected to minimize or suppress their own agency within the creative collective, their work was nonetheless defined by the visible product or effect it produced. Conversely, the very absence of visible impact or product defined women's production work. Indeed, success in women's crew jobs—such as script continuity—was measured by how little attention their work drew to itself (for example, lack of continuity errors). Inconspicuousness also characterized feminized labor outside of the film industry.[17] However, in that it concerns not only economic values, but also creative and artistic ones, examining the hiddenness of the female movie worker's role adds dimension to the historical erasure of women's work.

And yet, the fact that so many expectations for women's role performance were implicit makes it difficult to account for them; moreover, although gendered expectations were universal, women interpreted their roles individually. Secretarial roles at studios were particularly labile, as is any boss–secretary relationship—and made even more so by the industry's endemic cult of personality. Still, in order to ascribe credit, this hidden labor must be rendered measurable, and objective. To that end, I classify their form of feminized service to creative movie makers as "creative service," a concept devised to connect feminized duties that characterize so many women's professions to the oft-overlooked forms of feminine performance and emotional labor that were also deployed to support the movie maker or creative collective, to facilitate their creative vision and, by indirect means, to affect it. Creative service at its most basic shares the same prime directive ascribed to secretarial work in that the creative service worker "takes on all routine and some non-routine tasks that belong to the manager's role, thus freeing him/her to concentrate on core activities."[18] This guiding principle of secretarial work, applied to the creative process, bridges the rationalized clerical labor that made up the studio secretary's visible duties and the feminized service labor that was part of the job's hidden, unofficial requirements.

In creative service areas with which the rest of this book is concerned, subtle service characteristics are typically fused to more clearly observable feminized labor characteristics, often packaged together and delivered via feminine performance. Workers in such roles aid the creative process by serving as a repository for all its unappealing tasks, details, and emotions. This didn't just happen in the offices of movie makers. Many women's professions in film production required practitioners to adopt roles not as workplace wives or mothers to individual men, but rather as rule minders or taskmasters for entire productions, where they were charged with overseeing and maintaining the space of creative play—without stepping into it themselves. The following case studies, then, are offered to establish creative service as a provisional framework for understanding the larger purpose women's work at studios, connecting a seemingly disparate array of practices across a group of widely varying secretarial roles, all cohering around their most essential shared function: serving creative work by subtracting all noncreative work from the process.

Case Studies: Studio Secretaries and "Creative Service" to Movie Makers

According to Leo Rosten, movie makers were defined by their control over "the content and the implications of films"—their ranks comprised of an inner circle of *movie elite*, "the ones at the center of the power and prestige."[19] In the studio era, the secretary operated as the primary link between an exalted boss and his employees, situating her nearer to the creative process than nearly any other movie worker. Hundreds—perhaps thousands—of secretaries created their own versions of creative service to gratify the needs of high-powered employers. Still, the most successful secretaries used a similar mix of skills, hybridizing clerical/administrative with emotional/service labor, freeing the movie maker for creative work by acting as his emotional manager and solution shaper. These services were typically delivered nonthreateningly through a relatively narrow range of feminine roles (that is, office wife, den mother, gal pal/girl Friday, or good daughter).

The accounts that follow evince the common thread of creative service that ran through all the secretaries' jobs. These women are notable because they left behind descriptions of their experiences in their own words in the form of letters, memoirs, studio newsletter columns, and articles in other publications. The secretaries examined were also some of the most successful in their fields, excelling in positions at the highest levels possible within the realm of the movie elite. They used a diverse range of work styles in the creative service of a range of movie makers—from directors to studio bosses. Finally, they worked at such high levels and in such close proximity to the creative process that their experiences throw into greater relief the agency they possessed and the creative capital they wielded from their contiguous positions of power. Not every secretary or assistant worked for employers of such stature, so not every boss offloaded tasks of the same magnitude as Louis B. Mayer or Alfred Hitchcock did—and not every secretary's creative service can be linked so immediately to specific films or studio practices. At the same time, however exceptional the creative service rendered in these examples were, the service itself was typical, thereby illustrating a workplace dynamic that could be found in offices from the A-list to the D-list.

Valeria Beletti, Secretary to Samuel Goldwyn

Valeria Beletti's writings—in letters to her friend Irma Prina—span her time on the desk of Sam Goldwyn and several subsequent employers, from 1925 through 1929. Because they are related "in the moment" (or within a few days of the events she describes), her descriptions are quite detailed as to the who, what, and where of her day-to-day work. Cari Beauchamp, who edited the letters for publication in book form, corroborates many of these details in brief the passages that frame the collected writings. Though Beletti's career's impact on the movies is perhaps least significant of the women profiled here, her account provides a baseline of the traits and skills used in creative support of a movie maker. Moreover, her frankly related thoughts and feelings as a young female newcomer to the film industry complicate studio culture's representations of female workers as happily subservient helpers of men.

Although she was new to the movie business and still considered herself a "terrible typist" at the time she was hired into Samuel Goldwyn's office in 1925, Beletti had ten years' experience, having started at the age of sixteen in a New York patent office.[20] She worked as a private and social secretary to Goldwyn for a year and a half, becoming a valued employee.[21] A letter of recommendation from Goldwyn himself reads, "I consider her the best secretary I've had in 15 years."[22] That she so distinguished herself to Goldwyn indicates proficiency beyond straight clerical work in several important areas of creative service.

First, Beletti displayed the mixture of judgment and timing that allowed her to master the juggling act that was Goldwyn's office. In the computer age, this work mode is understood as multitasking; however, when Beletti wrote of her job for the first time on February 19, 1925, she simply listed the variety of ongoing processes she was required to track concurrently, keeping each in mind as she moved between them:

> As Mr. Goldwyn's Secretary I come in contact with every phase of the movie industry; looking for new material; keeping in touch with producers in New York; reading new books; turning over possible material to the scenario writer who happens to be Frances Marion; hiring actors and actresses, directors, camera men; keeping in touch with the art director, publicity man, the projection and cutting room and ever so many other things.[23]

Beletti also took on extra work from director Henry King and others in Goldwyn's organization who had no secretaries of their own.[24] If she were permitted to refuse this additional work, she never reported doing so. With three films in production, the strain was evident:

> I have to write nearly all of Mr. Goldwyn's letters because he certainly doesn't know anything about grammar. . . . My work is really hard. I have to talk, talk, talk all day long. People are constantly wanting to see Mr. Goldwyn about getting into pictures or they have "a marvelous story" that they can't mail to the office, but must see Mr. Goldwyn personally about it. I have to smooth things over and keep them away.[25]

Clearly, Beletti acted as her employer's representative and served as the point of first contact in his office. Representing Goldwyn meant deciding which visitors, calls, and correspondence genuinely required his attention. Weighing the importance of incoming tasks or concerns and eliminating those beneath the movie maker's attention is another characteristic skill of creative service. Secretaries describe themselves as acting as gatekeeper or filter. In her time with Goldwyn, Beletti filtered both the paper and the people that came through his office, freeing up Goldwyn to focus on the upper-level executive and creative decision making that made up the majority of his job. Even when serving as interim secretary for King, or for actor Ronald Coleman, she was doing so for the benefit of Goldwyn, so that his director and star would be free to engage with him on higher-stakes matters.

Beletti also relieved Goldwyn of errands and other tasks related to his private life. These tasks often required characteristics of emotional labor and service, such as tact, discretion, and a gingerly managing style. Not just any employee could be counted on to keep the personal information revealed by such errands in confidence and to maintain a neutral countenance during and after personal requests. Indeed, although glibly referred to as gossips, female employees were good candidates for keeping confidences, another characteristic trait of secretaries' creative service, in part because there was little danger of their promotion beyond the level of secretary, and thus little chance that an employer might one day engage as an equal with a former employee with stored knowledge of his personal problems and proclivities. All such qualities were required for daily dealings with Goldwyn's wife and home life. Serving as confidante in these smaller, personal matters qualified her for other levels of trust, such as purchasing bootlegged liquor for a company party.[26]

Creative service to Goldwyn required Beletti to winnow out tasks and information based on her perception of his emotional state, judging when and how to bring up potentially upsetting concerns, and gauging Goldwyn's moods to know when not to approach him with more routine tasks requiring his attention. She was also frequently called upon to manage her own emotions in response to Goldwyn's or to neutralize the emotional content of his messages to other employees. This set of skills—a form of emotion management—was crucial to the success of secretaries to movie makers. For Beletti, emotion management was especially important and especially difficult, as Goldwyn's temper was notorious. She reported being "bawled out" a number of times while still learning the ropes at her job.[27] Months into her employment in his office, she complained that Goldwyn was "getting awful— I was on the verge of quitting a few days ago. . . . I'm supposed to know what he's thinking about without his telling me. However, as my clairvoyant powers are nil, I just can't do it."[28] Beletti soon learned how to interpret his silence, anger, and obliviousness, and how not to take it personally.

Beletti, like other successful secretaries, dealt with Goldwyn's temperament and idiosyncrasies not by discussing her concerns with him or by negotiating a compromise, but by adjusting to his needs and absorbing any emotional backlash they produced. She acknowledged the toll that this part of her job took on her near the end of her tenure for Goldwyn: "Perhaps you know what it means to work for people who are very temperamental. The work is not heavy, but it is just the strain of not knowing where you are most of the time."[29] Of course, in handling these challenges cheerfully and without complaint, Beletti made herself indispensable, which, true to the nature of creative service, made more work for her in service of Goldwyn's interests and at the expense of her own. Though the job took its toll on her, she was rewarded for her successful, intelligent managing of Goldwyn's life with Goldwyn's respect for her judgment. He signaled as much when he asked Beletti's opinion, reviewing a cut of *Stella Dallas* with her and Frances Marion: "I told him what I thought, so Mr. Goldwyn wants me to come in with him tomorrow and we are going through the picture again and see what can be done about it."[30] When Beletti left Goldwyn's employ to travel abroad, Marion offered, on her return, to take her on as a secretary to help her develop as a writer.[31] This proposal was not surprising given Marion's characteristic mentorship of female underlings. But the normally terse Goldwyn also pledged support for Beletti, offering to buy any story material that she might come across in Italy.[32]

Perhaps most significantly, Beletti was able—through her standing with Goldwyn—to influence him in way that would prove important, both for Goldwyn and for aspiring actor Gary Cooper. Beletti describes her role in Cooper's early career in July 1926: "I raved so much about him to Mr. Goldwyn, Mrs. Goldwyn and Frances Marion and our casting agent—and in fact anyone who would listen to me—that Mr. Goldwyn finally wired to camp and asked our manager to sign him under a five year contract."[33] Though many would later claim credit for discovering Cooper, Beauchamp points out that "almost everyone's versions contain a mention of his being 'the boyfriend' of Goldwyn's secretary as the way he came to their attention."[34]

Marcella Rabwin, Secretary/Executive
Assistant to David O. Selznick

If Valeria Beletti represented the baseline of creative service, Marcella Rabwin's work as executive assistant to David Selznick, which spanned roughly a decade between 1932 and 1944, demonstrates the potential creative impact of such a role. Rabwin worked from a similar toolkit: multitasking, filtering, serving as confidante, and managing emotions. However, during years with Selznick, she extended the role of administrative "work wife" to the creative process, helping him stay the course in production on his films. Rabwin wrote about her experiences in her memoir, *Yes, Mr. Selznick: Recollections of Hollywood's Golden Era*, devoting most of the book's pages to Selznick and the other notable figures she encountered throughout her career, but saving the final chapter for her own professional journey. Though her writings add a sense of texture and specificity to what work for a producer like Selznick entailed, her memories, related fifty years after the fact, don't always match other historical accounts of the people and productions involved. In one significant example of the "fuzziness" of her memory, Rabwin accuses Clark Gable and his "good friend" Victor Fleming of racism and poor treatment of black actors on *Gone with the Wind*. Other accounts of that production attribute all the prejudice to Fleming and assert that Gable threatened to quit if Fleming didn't integrate the studio bathrooms, which had been labeled with "whites" and "colored" signs at the start of production.[35] It is impossible to say definitively what happened in this particular incident. What seems clear, both in this instance and in Rabwin's other mentions of Gable, is that Rabwin strongly disliked Gable, and that her personal feelings about him may have resulted in distortions. Though her feelings about most of the other figures she recounts in the book were neutral to positive, it often seems that the way she felt about certain people is guiding her memories of them.

Rabwin clearly loved Selznick far more than she hated Gable, and so her accounts of him may be subject to similar distortions and exaggerations. More striking than her glossing over of Selznick's flaws, such as his sexual advances on other secretaries, which she admits to overlooking, is the sense, reading her memoir and her other writings from later in her life, that the author is attempting to claim a greater share of the credit than she received in other histories for being Selznick's right hand. Unlike the inconsistencies in the Gable story, which mostly muddy the waters around the actor, this minor legacy burnishing provides important evidence. If Rabwin's memory was shaped by years of seeing herself fall further and further into the margins of Selznick's story, then her "exaggeration" of her role might better be understood as a form of subjective truth that could only be produced by her reflection on her career at end of her life. Rabwin *was* incredibly important to Selznick, and so, on that score, her tendency to round up the number of years she worked for Selznick may be an embellishment, but one that reflects larger issues of creative credit (or lack thereof) for secretaries or assistants of her stature.[36] The tension between her importance at the time and her need to reassert that importance later in life is significant for this case study. Her feelings, looking back on the work in 1999, provide a kind of balance for accounts like Mervyn LeRoy's, which paint secretaries like Rabwin as selflessly giving up their lives for the men they serve without regret.

Beautiful enough to have been screen tested at MGM, Rabwin (née Bannett) graduated from UCLA in 1928. She found her way into the film industry after the head of dressmaking at the department store where she worked was invited to take over a similar department at Warner Bros. Rabwin joined her there until she caught the attention of a producer, who asked, "What are you doing buried back here? . . . You belong up in front in the secretarial department." She recalled, "I rented a $5 a month typewriter and borrowed a Gregg shorthand manual from the library and planted myself at my kitchen table for the weekend."[37]

This preparation led to brief stints as secretary to Arthur Caesar, Darryl Zanuck, George Volck, and later work as an agent under Myron Selznick. She became Selznick's top earner, at which point, Rabwin recalled, male agents complained: "Their pride was adversely affected."[38] When faced with a salary cut to appease the agents' egos, Rabwin elected to "continue my education in film production" by moving on. She next worked at RKO as secretary to studio head Abe Schnitzer, but in 1932, after she had spent three months with him, the studio was bought out and its executives fired. "That meant me, too," wrote Rabwin, "except that I refused to leave." A friend who worked security continued to let her into the studio, where, for three weeks, "I took refuge in the closet-like teletype office . . . back pressed up against the door, my chair squeezed into the small space between the teletype machine and the exit." As Rabwin recounts it, these unpaid efforts were finally rewarded when David Selznick, who had come to head the studio, stopped her in the hallway. "What's your name? Where do you work?," he asked, before hiring her on the spot. In her tenure with Selznick, Rabwin worked her way up from secretary to executive assistant with several secretaries working underneath her and earning "the rarest of lofty feminine salaries."[39]

In many senses, Rabwin was a pseudo-executive, yet her role was still closer to that of secretary. When studio bosses appointed male executive assistants (for example, Walter McEwan at Warner Bros.), the men did not sit sentinel outside of the boss's office (often, they had their own) or exhibit many of the other signs of secretarial creative service. Male secretaries and executive assistants functioned much as they did prior to the field's feminization. As Lynn Peril explains, secretarial manuals asserted that "the *'higher type'* of office hired 'wide-awake' men to act as secretaries/junior executives, while female typists and stenos, possessed of little more ambition than to marry and leave the workforce as quickly as possible, pounded out the drudge work."[40] Male assistants were similarly distinguished from female assistants in the hierarchy, particularly through the amount of routine and service work they did for their employers. While female assistants were viewed as glorified secretaries, male assistants were understood as executives' apprentices with advancement potential.[41]

Rabwin's success was no surprise to anyone who knew her. In 1936, Charles Boyer named her "one of 10 most interesting and beautiful women in Hollywood."[42] Her longtime friend Lucie Arnaz wrote of her, "This was the first woman I ever knew who really read books! Voraciously! Who could hold court on any subject with anyone on the planet, king or Culligan Man."[43] Though she had left her short career as a literary agent behind, Rabwin claimed to have launched Ayn Rand's screenwriting career, recognizing the talents of the young writer, who at the time worked in the RKO costume department, and convincing an agent to pitch her writing to studios. The agent agreed, but it took Rabwin's continual follow-up calls over the next few weeks before the he finally got an offer from Universal. Rand quit her costuming job the next morning and began work on a novel. In true Randian fashion (at least as Rabwin tells the story), Ayn never thanked her.[44]

Selznick biographer Bob Thomas described Rabwin's "uncommon faculties," saying, "She could distribute and file the memos he broadsided to all important workers in the studio. Her sense of order brought organization to his work day, which otherwise might have fallen into chaos. She shielded him from time-wasting appointments and relieved him of routine functions."[45] As Rabwin rose in the ranks of Selznick's office, her attention was increasingly focused on filtering. Her job was to do "everything that Selznick did not have the time or the inclination for."[46] She "anticipated his every need" on a moment-to-moment basis and, in a more global sense, administered his office and, in later years, ran Selznick's own company, Selznick International Pictures.[47] Rabwin describes wide-ranging tasks in the following passage:

> I ran David Selznick's office, bought his underwear, executed his business orders, ran films for him all night long when he was nervously awaiting the birth of his son. I have rushed to an emergency sneak preview at the 11th hour to take notes when someone else couldn't, leaving guests at my dinner table. I chauffeured him served

his ambitions, loved his family, kept him in cigarettes, assuaged his disgruntled appointments, and placated his bookie. I wrote most of his business letters and many of those famous memos, and signed his DOS as authentically as he could. I sat through Hattie Carnegie fashion shows when he and Irene were shopping. I kept some people from seeing him, arranged for others to do so, and relieved him of some myself. I forced food on him, like a Jewish mother. . . . I arranged meetings. I cancelled meetings. I held meetings. And, as he wrote on that dedication copy of the script of *Little Lord Fauntleroy*, I shared his dream.[48]

This passage demonstrates the variety of responsibilities required of most secretaries and assistants at this level—but also provides a sense of the intimacy between Rabwin and Selznick, as well as his wife, Irene Mayer Selznick. Rabwin also interacted with Selznick's mistresses: "I had very subtle ways of handling his personal life. I was involved in it and I had to accept some responsibility. So after an evening out, I would forget to send the flowers. I would forget to send any kind of gift. So I stopped a great many of the affairs midstream. Because I loved Irene. I wanted to protect her. And after I left I couldn't protect her anymore. And she was cast aside for Jennifer [Jones]."[49]

Any divorce is too complex to be explained by one cause. However, by many accounts, Rabwin's role in Selznick's working life was keeping his excesses—his use of stimulants, his fits of temper—from derailing his projects. Thomas explains that, along with Irene Mayer Selznick and production manager Ray Klune, Rabwin was one of the stabilizing influences that had helped discipline Selznick's erratic talent, adding, "She smoothed over disputes between David and employees, delaying the dispatch of angry memos until his temper had subsided. More than once she had snipped off infatuations by failing to transmit messages."[50] Given how divorces can play havoc with the working lives of the people going through them—the inherent emotional strain, the loss of home support—it isn't difficult to imagine that the reverse might also be true: that losing Rabwin as his stalwart "office wife" might have had a magnifying effect on Selznick's domestic troubles. And, after his split from Irene, Selznick's career unraveled apace with his personal life. The loss of Irene—the producer's forward-thinking, creative compass—could be detected in his subsequent work. Selznick's *Duel in the Sun*, which starred Jones, was a disaster, financially and artistically. Meanwhile, Irene went on to produce Tennessee Williams's groundbreaking *A Streetcar Named Desire* on Broadway.[51]

Rabwin was deeply involved in the lives of other members of the Selznick family, from his "brain-damaged brother, Howard," whom she kept out of the papers, to his father-in-law and sometime boss, L. B. Mayer, with whom she served as intermediary, "often patching things up between employee and employer."[52] She also oversaw "that purgatory known as 'the outer office,'" where employees waited for Selznick for hours, but also stood in as his representative, running meetings and making decisions in his stead.[53] She was careful to make clear in later recollections that she was not giving her own orders, but "following up and carrying out" Selznick's when serving in this capacity.[54] Still, accounts indicate that she knew what those orders were without their having to be given, and could thus be seen as the producer's proxy.

Proxying, the advanced form of filtering Rabwin practiced when she stepped into Selznick's shoes, is another common part of creative service at high levels—when workers stand in for employers not just as symbolic representatives, but also as decision makers. Rabwin's daily work put her "in frequent contact with every department head and every director" and in "dealings with cast and crew." A profile of Rabwin in *Variety* explained that she kept Selznick's day free from detailed distractions by knowing "what he wants done, and how and when he wants it."[55] Rabwin was not the only woman in whom Selznick vested this sort of authority. Director Robert Parrish described how Jane Loring, who had been "a script girl, then a film editor, and was now the assistant to the head of the studio" was sent on location "as his representative. She was the front-office boss, and a very capable one."[56] There was also Kay Brown, who served as Selznick's East Coast story editor, working closely with Rabwin. Rabwin described Brown as "a brilliant woman," responsible for Selznick's purchase of the book *Gone with the Wind*, and "more important to the operation than her title would indicate."[57]

Rabwin performed as an emotional proxy for Selznick as well. She explained the significance of such work in production, saying, "Perhaps my most important role was that of arbiter and mother-confessor when morale-destroying misunderstandings or clashes of temperaments threatened the quality of the film." When she wasn't managing emotions around the producer while accompanying him to a set, she was reshaping his words to contain their risk, editing his angry memos "so the message didn't make an enemy of the recipient."[58] She called Selznick "a very kind man, basically," but said that, "in the stress of production he became a different human being . . . a very cruel boss because he didn't hesitate to speak his piece."[59] As such, Rabwin explained, "It became part of my responsibilities to cajole, encourage, instruct, pamper and listen to our stars."[60] Rabwin served as Vivian Leigh's emissary to Selznick when the actress fumed that he must stop writing so many kissing scenes into *Gone with the Wind* because "I can't bear to kiss Clark—he smells so terrible"—a statement referring to costar Clark Gable's "odiferous" false teeth. "I delivered the angry message," Rabwin recalled, "and if you remember the film, you will realize how few the intimacies between Scarlett and Rhett."[61] She similarly acted as go-between for Selznick and his contract director Alfred Hitchcock during their troubled collaboration.[62]

As emotional proxy, Rabwin absorbed the anger of employees, perhaps, most notably, managing the general tension around production of *Gone with the Wind*, when Selznick's antics and angry memos threatened completion of the film. At MGM (where it was being shot), Rabwin recalled, "Everyone was upset. It was an upset studio."[63] She managed the fallout from Selznick's firing of director George Cukor (whom she called Selznick's closest friend) from the same film, and said she attempted throughout production and the rest of her tenure with Selznick to patch up their relationship.[64]

Rabwin also negotiated the immediate emotional climate in which she and Selznick operated, acting as psychotherapist and superego to his id. Their connection was such that many said Marcella was in love with David. She sounded a bit like a woman in love when she excused his inconsiderateness, claiming it "was compensated [for] by his geniality and brilliance." She added, "I cannot apotheosize him for he was too flawed. Still, I never lost my conviction that he had made an art of the cinema and was, himself, its greatest artist."[65] For Rabwin, the relationship more closely resembled that of a father and daughter. She recalled that "he treated me like a little girl," citing both Selznick's fatherly concern and Irene Mayer Selznick's motherly gestures. But by her own admission, there was some truth to rumors of her affection for her boss: "It is not possible to work side by side for so many hours of the day, for so many years, with someone you don't admire—and admiration is akin to love. Yes, I loved Mr. Selznick—adored him. There was something worshipful in my attitude otherwise I could not have spent so many overtime hours serving as his acolyte on the altar of cinema. I loved my work. I loved my indispensability to the most brilliant and talented film producer the industry has ever known."[66] That Rabwin was such a believer in Selznick the artist seems to have had much to do with her success in Selznick's operation, as her faith allowed her to make the same sacrifices Beletti did for Goldwyn, and indeed far more. In exchange for her devotion (and sacrifice), Rabwin garnered a particular kind of agency, administering Selznick's offices, serving as his proxy as well as his creative partner, and experiencing his triumphs and defeats by his side—a creative "work wife," complete with a wife's adoration. But the rapport Rabwin enjoyed with Selznick was by no means shared by every Selznick secretary. Like Sylvia Schulman, whose fictionalized account of the producer in her book *I Lost My Girlish Laughter*, scathingly depicts him as careless, hot tempered, and sexually harassing, Lois Hamby, Selznick's

secretary in the 1950s, recalled fending off advances from the boss, saying, "He made passes at me. But the first pass he made at me I said 'What's
this? Knock it off!' And he says 'What are you doing, you saving it up for
college?' I said 'No. If I want it I can get it but just forget it. Forget it.' He
looked at me, laughed and said 'Take a memo.' and that was it."[67]

There is little record of Rabwin's making creative decisions that affected
films like *Gone with the Wind*. Yet she was so central to Selznick's creative
life that they were often treated as a unit, with stars approaching Rabwin
for successful results rather than approaching Selznick directly. She also
contributed to Selznick's creative work in the sense that she filtered noncreative business away from him. One major example of this occurred
when, as secretary, she supervised the installation of Selznick International
Pictures in the former Ince Studios, freeing Selznick to work on the script
for *Little Lord Fauntleroy* (1936), which would be his first film there.[68] Such
incidents give an indication of the impact Rabwin made on Selznick's creative life. Yet, like many women in the studio system, she achieved her professional stature in part through willingness to confine the signs of her own
authorship to such places as the reference initials at the bottom of Selznick's
memos. However, in later years, she seemed to desire further recognition
and acknowledgment for her work, becoming a popular speaker at *Gone
with the Wind* conferences, penning her memoir, and in general asserting
her authority as Selznick's right hand.[69]

Rabwin's impact, via her covert form of agency, was not without its costs.
Over a decade with Selznick, her sacrifices accumulated. Unable to take her
influence and power-by-proxy out on the open market to obtain a job more
befitting her experience and skills (say, as a producer), Rabwin was trapped
with the chronically inconsiderate Selznick, who loved her, but only when
he stopped thinking about his own concerns long enough to do so. Rabwin
predicted his every emotion; and yet, hunkered over his latest project, he
forgot her existence. This detachment came to a head one night, in 1944,
when Selznick delayed Rabwin's leaving the office for a dinner party she
had discussed with him in advance. The appointed hour rolled by with
Selznick locked in his office, responding repeatedly, "Another five minutes
and I'll let you go." Rabwin was forced to confront her untenable situation:

> My husband was fuming and I was twiddling my pen and cringing
> on the inside from the humiliation I was heaping on him. He knew
> always that my having attained the position of indispensable assis-
> tant to the film industry's most prestigious producer was a source of
> satisfaction almost equal to my happiness in my marriage. But there
> had to be a limit. This was the breaking point. He issued me an ulti-
> matum: "Make your choice." I made my final call on the intercom.

"I just want you to know that I'm leaving now." "Wait just another five minutes and I'll get to you." "I can't wait, Mr. Selznick. I'm leaving." "What do you mean you're leaving?" "I'm leaving for good." There was a little nervous laugh. "Are you serious?" Then, for the last time, as I had said thousands of times in the past decade, and almost sobbing, I replied, "Yes, Mr. Selznick."[70]

In this, her recollection of the events so many years later, Rabwin's much-loved career ended in a choice between her roles as a work wife and as an actual one.[71] Selznick had credited Rabwin with sharing his dream, but that credit was given privately, arbitrarily, and selectively. Her work wasn't her own without his say-so, and her power wasn't power outside of his sphere. When she could no longer make the intense personal sacrifices required to retain her agency and control, her work and its impact—outside of her own accounts and those of the people who would vouch for what she did— effectively ceased to exist. Marcella Rabwin is a credit to film history and to women's history. The example of her career highlights the nature of creative power and how tenuous women's connection to it was at this time, pre- cisely because it so often came through feminized channels.

Ida Koverman, Executive Secretary to Louis B. Mayer

There is no memoir or biography devoted to Ida Koverman and her work for Louis B. Mayer from 1928 to 1951, though she was among the most well-known and powerful studio secretaries. However, Koverman wrote columns for the MGM newsletter, and many major players remember her in their biographies. If Rabwin exemplifies creative partnership (without equal benefits) through creative service to a producer, Koverman's career demonstrates the covert managerial partnership between trusted secretar- ies and their front office employers. And, although some who worked with her claimed she "damn near ran the studio," Koverman's case also shows the limits of that power, coming as it did from her position as an extension of Mayer.[72]

Ida Koverman (née Brockway) met Louis B. Mayer in 1928 while work- ing as executive secretary for Herbert Hoover's presidential campaign (as she had for Calvin Coolidge in 1924).[73] Koverman "had stars in her eyes and wanted to work in Hollywood" and "was anxious to leave the East Coast" for personal reasons.[74] As the first woman secretary of the Republican Central Committee, Koverman could help Mayer advance in the Republican Party through her "close friendship with Mr. Hoover and others."[75] Mayer hired her as his executive secretary (sometimes referred to as his "assistant"), a capacity in which she served until he left the studio in

1951, when Dore Schary promoted her to director of public relations.[76] In Mayer's employ, she was considered both secretary and pseudo-executive, a well-known figure on the lot, synonymous with Mayer in the minds of many of the executives and emblematic of Mayer's studio brand to many workers.[77]

"The Women Who Run the Men," a 1942 *Variety* feature, profiled those women whom the trade paper considered the six most important secretaries in Hollywood, invoking a football team's offense in characterizing them as an "all-American, six-woman line, operating up ahead of the ball while clearing away debris, the trivia, the unnecessary and the unwelcome—the buffers before whom Fate itself might recoil." Koverman topped the list in a section called "It's This Way, Mr. Mayer," which describes her as a "confidential advisor and buffetteer in chief," who "knows what L. B. Mayer is going to do with his day before he does." [78] Koverman was familiar to anyone who wanted to see Mayer. With the help of two other secretaries, she served as his gatekeeper and made it "very difficult to see [him]" according to Lillian Gish, who described Koverman as "one of the most formidable women I've ever seen in my life. She virtually ran Hollywood."[79] Some disliked her for her forbidding presence as Mayer's gatekeeper. Evie Johnson (wife of actor Van Johnson) called her "Mount Ida. . . . She was Margaret Thatcher, but much worse. She was dictatorial, unbending, and unyielding . . . not a nice old grandmother."[80]

Yet most accounts of Koverman are warm and admiring. Colleagues respected her as an extension of Mayer and an architect of many solutions to administrative problems. MGM executive Robert Vogel called Koverman "brilliant," explaining, "In the morning at nine o'clock, there'd be a hundred people there to see Louis Mayer and she would talk to each one of them, very briefly, and say, 'That's for Mr. Mannix. Go and see Mr. Mannix.' 'That's for Mr. Shearer. Go and see Mr. Shearer.' 'Mr. Mayer will be able to see you on Friday.' And so on and so on, the whole damn thing. I found out that everybody said they had to see Mr. Mayer right away."[81]

In politics, Koverman outranked even Mayer. Her prior credentials were significant, and she continued to work in political organizations and serve at high levels on committees for the party and its conventions.[82] Separately, and in Mayer's service, she was "an adroit politician, public speaker, lobbyist, and expediter." And with his executive secretary's "ready access to the innermost GOP circles," Mayer's fortunes in the party rose considerably.[83] During Republican presidencies, "no Hollywood figure was a more frequent visitor to the White House than Ida Koverman."[84] When asked to run for Congress, however, Koverman refused.[85]

Despite her influence and power in her own right (and as Mayer's proxy), Koverman relied on emotional labor and feminine performance to package her authority. Like other secretaries and assistants, she managed emotions—from Mayer down to his underlings. And because Koverman was "less than overwhelmed by the ego trips of stars," she was often called to handle temperamental employees.[86] One such employee was Mayer himself. Cari Beauchamp explains, "For Frances [Marion], Ida's presence meant a friend at the door; someone to gauge the boss's mood and a fount of information on the latest comings and goings. She liked and respected Ida but she never understood her unending patience with the man Frances had decided was 'a pompous ass.'"[87] Koverman is also notable for the feminine role she played for an entire studio—as mother figure in what L. B. Mayer liked to call his MGM "family." In many ways, Ida was the lot's matriarch. As the MGM newsletter described her, "Interested in you and me, a constructive critic and therefore always helpful . . . always, too, sincere, zealous, unselfish . . . long the able executive secretary to our own Louis Burt Mayer . . . and one of our Club's staunchest personalities."[88] In her forties, when she first came to MGM, with gray hair and a matronly appearance, she "brought a dignity and class to the studio," and was "treated like the 'queen mother.'"[89] Koverman shared Mayer's tastes and values and, in her stern-yet-caring role, often disseminated and enforced them in the rest of the studio. According to Frances Marion, "Ida had become a mother image to all the fledglings away from home, while the disobedient young were afraid of her because of her affiliation with Mayer."[90] Likewise, actor George Murphy said of Koverman, "In her wonderful way she went through life sharing an unbounded motherly devotion, a sort of patron sainthood . . . always ready to help with her great wisdom and understanding, indulgent but stern, ruling her big family with a mother's pride."[91] Koverman took stars to task "if they started believing their publicity," helped evolve dress and behavior codes for the studio's actors, and was responsible for the studio's talent school, as well as the Little Green Room, a play area created for the studio's child actors, staffed by a "studio mother."[92]

Koverman was also a mother figure to Mayer himself, playing "official studio hostess" at the many sumptuous MGM parties—not to mention the homey "surprise" birthday party held each Fourth of July for all employees in the studio commissary.[93] It was to Ida that L. B. turned "if there was a question about which glass to raise or how to address a visiting ambassador," and for advice in matters of etiquette "both at his home and at the studio, where she became mistress of protocol, supervising elaborate luncheons."[94] She drew upon her experience with Hoover to give Mayer "much of the polish he acquired with the years," and got him to quit his habit of digressing during his speeches.[95] In private dealings, Mayer "depended upon her," and she, in turn, "guided him with a firm, matriarchal hand."[96] She was able to do all of this by managing her own status relative to Mayer's—advising him in politics, offering constructive criticism of his mode of address "without ever challenging Mayer's eminence as a public spokesman," and exercising her power "with no apparent interest in its outward trappings."[97]

Koverman's motherly guidance made her, arguably, the most powerful influence in Mayer's professional sphere, which—in turn—made her, in the words of Beth Day, "one of the most influential women on the West Coast."[98] Off the lot, she helped launch Hedda Hopper's career as a gossip columnist in 1938. In its first two weeks, Hopper's column had drawn only mild amusement from Hollywood—until Koverman reportedly remarked, "They've laughed long enough. It's time they laugh out of the other sides of their mouths," and threw a party that Hopper recounted as a "female 'who's who.' Judges, lawyers, doctors . . . and writers."[99] Sixteen papers soon picked up her column.[100]

Koverman used her power with Mayer and the authority it lent her in talent acquisition and development, acting as a scout for new stars while keeping an eye on those under contract. Unable to sign actors herself, she nonetheless had the bosses' ears. According to David Shipman, "Many of M-G-M's most successful stars had been brought there because of the enthusiasm of Ida Koverman."[101] A number of different accounts cite Koverman's role in bringing Judy Garland to Mayer and others' attention after Garland auditioned for Koverman and Roger Edens.[102] Though the events are different in nearly everyone's recollection, Koverman seemingly figures in all of them. As Edens tells it, "I called Ida, L. B. Mayer's secretary, she called Mr. Mayer, and he called the lawyers."[103] While dispassionate in his reaction to Garland, Mayer reportedly ordered contracts immediately.[104] The story of how Koverman later convinced Mayer to cast Garland when he had doubts about her demonstrates not only her influence over Mayer, but also the subtle means through which she obtained it. According to Frances Marion:

> Ida came to me in despair. "The Boss has lost interest in Judy. Whenever I suggest her name for a small part in a musical, all he says is, 'stop bleating! I'm running this studio, not you!'" Her lips were tight-pressed for a moment. "But I'm not giving up! I'll never give up! Somehow I'll manage to get his interest back to Judy again." . . . When a clever woman plots against a clever man, her only hope is to attack through his weakness. L. B. Mayer was a sentimentalist. Sentimental love stories, mother-love stories, and sentimental songs moved him to tears. But nothing touched him more deeply than the hauntingly melodic "Eli, Eli," which he had heard for the first time when it was sung by Bele Baker at the Hammerstein Music Hall. And this is what Ida Koverman had Judy Garland learn. And which she

sang one afternoon when he was alone and depressed. "You'll never leave our studio," said Mayer when she finished, the sob in his voice matching Judy's.[105]

Rather than addressing the matter of Garland's future head-on with Mayer—as one executive might approach another—Ida subtly influenced him by cannily choreographing his emotions to achieve her own creative ends. (Sadly, according to Marion, as a result of the same meeting, Mayer ordered Garland—an insomniac—be given sleeping pills and "a supply of that stuff they use to pep you up in the morning," as Ida's face "turned pale with anxiety.")[106]

Koverman also played champion in the case of Clark Gable, whose initial screen test for MGM Irving Thalberg deemed disastrous.[107] Like Frances Marion, Koverman saw Gable's potential and "animal grace" and refused to accept defeat when the male decision makers disagreed, reportedly exclaiming, "That's the trouble with this business. It's the men who pick the stars and the women who react to them."[108] She ran the same test for female employees and reported the results to Mayer: "It was a landslide vote in favor of the man with the big ears."[109]

Perhaps the greatest sign of Koverman's influence and authority was that so many executives took her advice. She became "known as 'the woman to see' to influence picture assignments."[110] Said *Variety*, "There's very little taking place on the lot that she hasn't exerted some influence upon. For hers is the unseen hand that pulls the strings that make the puppets dance and her name is spoken softly and with some degree of veneration."[111] Gradually, incidents like Gable's screen test helped Koverman gain authority as more than an extension of Mayer. Marion called her "the only woman executive whose advice was respected by the male stars."[112] Vogel echoed Murphy's claim that Koverman was "the greatest woman in the industry," saying, "She was everything. That's the point."[113]

And yet, outside of scattered anecdotes about Garland or Gable, Koverman's name is largely absent from film history. Her considerable impact on the industry and its films is inevitably attributed to Mayer. Though she was promoted to her own executive role after Mayer's retirement and continued there until her death in 1954, for much of her career, she straddled the line between a secretarial role and an executive one; doing so required that she not only have qualifications in the same areas as male executives but also show a willingness to serve as mother-nurturer to a mogul, his pals, and their studio.

Like Rabwin, Koverman downplayed her own role. In a studio newsletter feature on the work of the front office, she claimed that the men she served did not enjoy entertainment "as we do, sitting back in our seats completely relaxed," but were always on the job, looking for personalities in the films. She reminded the newsletter's readers of "the tremendous weight of responsibility which rests on the shoulders of what we call the 'front office'—the heavy task of keeping the wheels of this great studio constantly rolling so that the rest of us, numbering five or six thousand workers, may have good jobs."[114] Koverman, by the admission of many of the executives described in this passage, bore much of that responsibility herself.

Meta Carpenter, Secretary to Howard Hawks

Meta Carpenter served as secretary and script girl to Howard Hawks on and off throughout the 1930s and 1940s. Her memoir, *A Loving Gentleman: The Love Story of William Faulkner and Meta Carpenter*, cowritten with Orin Borsten, details her work for Hawks, as well as her decades-long love affair with William Faulkner, whom Carpenter also assisted during his collaborations with Hawks and in the periods when she and the writer lived together during his stints in Los Angeles. Her own words paint a general picture of secretarial work in service of a director—a lower-status role than that of the producer or executive within the studio system, but also a more straightforwardly creative one.

Carpenter's romance with Faulkner is significant to a discussion of her work in that assessing the Hawks–Carpenter relationship alongside the Faulkner–Carpenter relationship illustrates the extent to which the workplace dynamic between secretaries and movie makers resembled a romantic relationship, demonstrates as much as possible the lived experience of that dynamic, and evinces the possible ramifications of a creative workplace in which so many of the female employees participated in the creative process as a function of this sort of relationship.

Born Meta Doherty in 1907, Carpenter came to Los Angeles in the early 1930s with then-husband Billy Carpenter.[115] They lived apart (he in Santa Maria studying aeronautics, she in Los Angeles working a "meager-paying job at the Platt Music Company in Los Angeles") and divorced after realizing they were happier that way.[116] Keeping the name Carpenter, Meta found work as a secretary in casting at Columbia with the help of Katherine Strueby, Preston Sturges's secretary, with whom she resided at the Hollywood Studio Club. She worked briefly in the production office of cowboy star Buck Jones, and it was there that Howard Hawks hired her to replace a secretary who left his employ to have a baby. When Hawks moved to Samuel Goldwyn Studio to work on *Barbary Coast* (1935), Carpenter accompanied him.

As Hawks's only secretary, Carpenter's daily work included more typing and clerical labor than the jobs of Marcella Rabwin or Ida Koverman. And, at the time she came to Hawks's office, Carpenter recalled, she was light on just such skills. In her words:

> My shorthand, a mix of incomplete night school Gregg and arcane symbols, sufficed only because my employer dictated slowly, weighing each word to achieve the conclusion of thought on which he prided himself. I had only recently attained a degree of proficiency as a typist. When I wasn't being the magnolia-voiced, competent Miss Carpenter at my desk, I was major-domo to Hawks, superintending the moving of his office furniture and files (he was one of the first freelance directors) from studio to studio; seeing to it that his race horses were stabled, fed, and shipped to the proper tracks; helping [Hawks's wife] Athole Hawks with sundry matters at their home . . . and generally relieving him of tiresome tasks that he delegated to me.[117]

This work, as always a mixture of clerical and nonclerical, physical and emotional labor, was often more personal than professional, reflecting that directors at work on a project did less office work than producers or moguls because their managerial work took place on the set in production. Still, Carpenter functioned in the same capacity as other secretaries, freeing Hawks to think of nothing but his project and then go home, where his wife, with the help of Carpenter and other servants, would continue to facilitate this freedom.

In fact, some of Carpenter's duties were so personal that they seem closer to those of an actual wife than a work one. She recalled that after Irving Thalberg's death, "I telephoned long lists of relatives and friends, ordered groceries and meats, answered letters of condolence, and made endless family arrangements."[118] This incident demonstrates the support functions secretaries were often counted on to provide movie makers—both at work and at home—so they could stay on schedule and produce creative results. The need for such was even more pressing for directors in production, when large dollar amounts were attached to even the smallest of lags in time that could be created by turmoil in a director's home.

However, juggling a director's affairs during production came with benefits. For Carpenter, there was the opportunity to travel with Hawks and meet interesting people such as Amelia Earhart (with whom she flew over Los Angeles in a Stimson), Tallulah Bankhead, Maurice Chevalier, and others.[119] While preparing *Barbary Coast*, she and Hawks stayed at the St. Moritz Hotel in New York. She lived in what, by day, was the workroom for Hawks, Charles McArthur, and Ben Hecht.[120] For five months, the two writers dictated rapid-fire dialogue and plotting to Carpenter.[121]

Though any secretary to a director would likely have spent time on set during the production of one of his projects, Carpenter became part of the crew as script clerk on *Barbary Coast* when "Mr. Hawks asked if I could do both" for a salary increase from $35 to $42.50 per week. When Faulkner pointed out that Hawks saved money by having her do "double duty," Carpenter insisted that she benefitted from the arrangement as well in that "he made allowances for my mistakes and he had his own cutter, Eddie Curtis, work six weeks training me."[122]

Still, after the rush of excitement on the first film with Hawks, the arrangement wore on Carpenter. She often stayed late at the studio, typing her notes for the editor or taking dictation from Hawks.[123] Carpenter soon realized how these exertions—in a field where she had some prospects but was unsatisfied by the work she herself did—cost her creative ambitions as a concert pianist (for which she was classically trained), saying, "I couldn't be Howard Hawks' secretary/script girl and still have the time, the boundless energy, the discipline, the purity of spirit—the last above all my girl—to devote to music."[124] And, the salary she received in exchange for her efforts was barely enough to live on, even at the Studio Club, much less to foster her musical ambitions. ("Where would I find the money to pay a top teacher? I thought of my purse on the last day of the week, with never more than a crumpled dollar bill or two.")[125]

Creative service to a director in production was often psychologically draining as well. Carpenter's work for Hawks on *Come and Get It* (1936) was more stressful than the already "nerve-battering processes of moviemaking" because of conflicts between Hawks and producer Samuel Goldwyn. The result was that "from the moment my employer strode onto the set each morning, tensions began to build and layer until, by late afternoon, the air on the sound stage was thick with strain, indecision, and malevolence."[126] As Hawks's secretary, Carpenter was a party to this tension, and felt its impact. Though technically a member of the crew, even as script clerk, she was closely aligned with Hawks in the contentious space of production. In both roles, she owed her loyalty to Hawks, even when it conflicted with her own ambitions in script continuity. "Wanting in solid experience as a script girl," she made the best of things, squeezing "everything from my mind but the mimeographed pages of the screenplay."[127] But the time came when her employer's interests trumped her own. With one week left in production, Hawks quit the film. Carpenter recalled, "Hawks returned to the sound stage for long enough to pick up a few personal possessions from his trailer office, then walked off the set without explanation. Gathering up my marked script and papers, I followed, leaving actors and other company members mystified."[128] She had no choice in the matter: as Hawks's secretary and his hire as a script clerk, she was viewed by production as an extension of the director. Her crew identity, the skills she gained during production, and her personal qualifications seemingly meant nothing, evaporating once Hawks left the set. When production resumed under William Wyler, the production hired a script clerk who had worked with Wyler to finish the film. Hawks affected an unbothered, relieved air, leaving Carpenter behind to pack up his effects. Carpenter recalled feeling "crushed, a mere secretary once more, first obligations to Howard Hawks," explaining, "For weeks, I had been a member of a movie company, gaining authority in my work each day, and now, depressingly, it was all over for me but not the others."[129]

Carpenter's loyalty, which resulted in the "abrupt separation from the canvas chair on which my name had been imprinted and which had always been placed next to Hawks' leather chair on the set," did not entitle her to reciprocal loyalty. She left Hawks's employ as he was preparing *Bringing Up Baby* (1938), married German concert pianist Wolfgang Rebner, and moved with him to Europe. The couple returned to Los Angeles at the outbreak of World War II, but Rebner struggled to find work as a composer, and, despite her previous success, Carpenter fared little better. Desperate for work, she resorted to sneaking onto lots and knocking on doors of anyone who might someday need to hire a secretary or script girl. One door she didn't knock on was that of her former employer. Even though he always had a number of projects in the works, she could not force herself to call him, explaining, "My voice would quaver. I would stammer. I might weep. Howard was uncomfortable with desperate people. Better not."[130]

A dutiful secretary even after she ceased to work for Hawks, Carpenter absorbed this emotional and interpersonal discomfort rather than visit it on her former employer and chose instead to reestablish her career on her own. When she was later assigned as a script supervisor to Hawks's *To Have and Have Not* (1944), she gasped, "Did he ask for me?" and was told by the production manager that it was a routine assignment.

However distant Hawks was, Carpenter hardly makes him out to be a villain or even a bad boss. While still in Hawks's employ, she told Faulkner that she was "reasonably happy" with the director, adding, "He always treats me with respect."[131] She saw similarities between "Howard," to whom she acted as work wife, and "Bill," for whom she served as both secretary and romantic partner, reflecting that both were "moated men, closed off unto themselves."[132]

In many ways, Carpenter's creative service for Hawks was mirrored in her relationship with Faulkner. They fell in love after Hawks assigned her to transcribe Faulkner's handwritten pages of the screenplay for *Road to Glory* (1936).[133] In later writing sessions, Carpenter functioned mostly as a conduit, saying, "I made more than my usual percentage of typing errors in the letters and reports, but not in Faulkner's pages for *The Road to Glory*; they were the Talmud, the Bible, the Koran to me."[134] As their romance became physical, the work relationship did as well, and Faulkner, when out of ideas, would read over Carpenter's shoulder "brushing my hair with his lips."[135]

The blurred workplace boundaries didn't extend to the creative process, however. Carpenter once questioned Faulkner about a line of dialogue, "since as script girl, I would have to deal with screenplay inconsistencies," beginning, "Now Bill, in this scene you say—" before Faulkner interrupted that he didn't say anything: the character said it.[136] Here, true to the role of the secretary and script girl, Carpenter was concerned with ensuring continuity and clarity, both in production and on the screen. Faulkner, true to his role as creative, was unconcerned with such banal "details" as dialogue inconsistencies; he disavowed their existence altogether by insisting the characters were real (and therefore consistent). Carpenter was not without creative sensibilities and intuitions of her own. She played the role of rule-minder because that was her job—the creative service she was there to provide so that Faulkner could bring his characters to life undisturbed by particulars.

As their relationship progressed and Faulkner burrowed more deeply into his next novel, *Absalom! Absalom!,* and in later stints in Los Angeles as he worked on the manuscript for *The Wild Palms*, Carpenter's love for the romantic, impractical, antisocial, binge-drinking author took on greater dimensions of caretaking and required increased sacrifices on her part. In a letter to Carpenter written during a particularly stressful separation, Faulkner joked, "I weigh 129 pounds and I want to put it all on you."[137] Meta took on the weight of Bill's love, his creative frustration, his family squabbles, and his alcoholism. This dynamic continued intermittently throughout the years and Carpenter's two marriages to Rebner, during which, at least once, Faulkner nearly (seemingly, accidentally) killed himself in a fit of drunken despair. Neither Faulkner the romantic fool nor Faulkner the creative force, it seemed, could take care of Faulkner the man. Near the end of a stretch he spent Los Angeles with Carpenter in the mid-1940s, she resolved "not to let more than a few days go by without seeing him for at least an hour, to read his moods, to look for the danger signs. I could at least put a hand over the edge of a glass as a reminder of my concern."[138] Her own needs seemed to disappear in his presence.

Faulkner and Carpenter were lovers, not employer and employee. Her behavior toward him was that of a good wife, minus the wedding band. Her decision to accept Faulkner for all his faults was personal, as were the results of that position. And yet, their dynamic reverberates with the Hawks–Carpenter workplace dynamic, along with that of Rabwin and David Selznick and other secretaries and their employers. The disadvantages that resulted from this bond—the uneven ration of sacrifice to reward—were profound, indeed, reinforcing the already substantial barriers between a secretary's job and the junior-level executive positions nearest to it, and making her contributions even more difficult to distinguish from those of her employer. After all, per studio and larger culture, a loyal secretary lived to serve the man in her life and basked in his reflected glory, never seeking to rise to his level and claim some of the credit for herself.

Carpenter finally wrote about her connection with Faulkner twenty years after his death, and near the end of her own long career, during which she had endured innuendo from her peers and speculation by literary historians. Faulkner biographer Joel Williamson writes of her memoir, "There had long been a rumor of a Hollywood 'scriptgirl' with suggestions akin to those that go with travelling salesmen and working women," before concluding that Carpenter likely "felt that she had been pushed unfairly into obscurity" and set out to correct the record.[139] Carpenter herself explained the timing by saying that she had deluded herself into thinking that her affair with Faulkner was still the sort of small-scale gossip it had been in the 1930s, when the writer was less known. These illusions were shattered, she recalled, when she learned that

> unknown to me, the legend that William Faulkner was my lover had spread throughout Hollywood. . . . A number of film-studio writers, one or two from the group who knew us forty years ago, were only waiting until I died to turn out books about William Faulkner and his Hollywood script girl. When I recovered from the numbing shock, I made the decision that there was no option left to me but to write in collaboration with a close friend my own account of my years with Bill.[140]

In one sense, Carpenter wrote herself into Faulkner's life story, claiming due credit for the role she played as—most historians agree—the love of his life, and for sustaining him emotionally, physically, and intellectually during the creation of some of his great works, particularly the novels *Absalom! Absalom!* and *The Wild Palms*, and his credited and uncredited work on *The Road to Glory, Gunga Din*, and *To Have and Have Not*. However, it might be more accurate to say that Meta wrote Bill into *her history*, in that the memoir focuses on her own version of events, exposing Faulkner as deeply flawed and unromantically mortal at the same time that it expresses her love for him.[141] Carpenter also speculates, rather boldly, as to her impact on Hawks's creative life:

> Jean Arthur in all her variations, incapable of guile or artifice where her man was concerned, straight-shooting, accommodating,

undemanding, sweetheart and pal all in one. Clearly, Howard Hawks knew far more about his blonde secretary and her relationship with William Faulkner than I had deduced from his uninquisitive manner and masklike face. I make no pretense to having served as the model for the classic Hawks heroine, comfort and joy of the noble, stalwart Hawks Hero. The coincidence of timing and likeness, however, cannot be entirely ignored. If any part of me as I was then went into her creation, Hawks and the directors who borrowed from him are welcome to the bits and pieces.[142]

Though it may well be true, there is no evidence, other than this passage, that the "Hawks heroine" was based on her or her relationship with Faulkner. In truth, her description of Hawks's accommodating sweetheart sounds like many of the girls Friday present on the lots throughout the 1930s and 1940s but still absent from film history.

As much insight as it provides into Faulkner, *A Loving Gentleman* is very much Carpenter's story. Sections of the book show that her career, which was significant in the field of script supervision, went on without Hawks. Faulkner seemed as though he might die without her. When Carpenter learned of his death while on location with a film, she recalled sinking under the location's giant trees, feeling bereft, as a widow, adding, "I will never know how, an hour later, I was able to make my appearance at the set."[143] Yet she finished the day's work.

Peggy Robertson, Assistant to Alfred Hitchcock

Peggy Robertson's account of her work for Alfred Hitchcock, which spanned the 1950s through 1970s, comes in the form of an oral history collected by Barbara Hall.[144] Robertson first worked as Hitchcock's script supervisor and later as his assistant. Her importance to his work, along with that of the director's other unofficial collaborator—his wife, Alma Reville—have recently begun to receive more widespread recognition.[145] Though much of her work with Hitchcock took place after the studio era, Robertson's account is particularly illuminating in its description of how the director mentored her (in contrast to his treatment of his female stars), her clear impact on the auteur's films, and as the best example of a type of creative service that assistants to media makers commonly perform today.

Born Peggy Singer in 1917, Robertson met Hitchcock in 1949, when he returned to England to make *Under Capricorn* (1949).[146] "It's the only time I've ever asked for a job," she recalled. "I thought he was a wonderful moviemaker. So I went to the office at Granada Films, saw the production manger, and said I wanted to work with Hitch."[147] Once hired, she demonstrated her value to "Hitch," and he made her his assistant. Robertson followed him to the states with husband Douglas Robertson (a film editor) after the departure of his previous secretary, Joan Harrison.

Like Harrison had been, Robertson was accused of being Hitchcock's mistress. According to Hitchcock biographer John Russell Taylor, Robertson, "being the forthright, jolly-hockeysticks English lady she was," asked Hitchcock directly if "being his mistress like Joan was one of her duties." Hitch reportedly replied, "I can safely tell you that I was never between the sheets with Joan."[148] Whatever Hitchcock's behavior to other female employees, the relationship between him and Robertson remained one of collegial friendship. Robertson claimed he was a mentor to her and supported Harrison in her career as a producer, stating, "I *wasn't* the exception to the rule."[149]

Some of Robertson's regular duties mirror those of Koverman or Rabwin. Hitchcock disliked disagreements and also used the public perception of him as enigmatic to his professional advantage. Robertson often delivered messages for the director, especially on set, and smoothed out difficulties between Hitchcock and other associates. She consoled and bolstered leading ladies—among them, Ingrid Bergman, Jane Wyman, and Tippi Hedren.[150] She also took copious notes throughout production, even when not working as script supervisor, amended the record in the case of his exaggerations and memory lapses, and contributed to daily production reports.[151] Hitchcock secretaries such as Suzanne Gaultier supported Robertson, maintaining Hitchcock's office while he and Peggy were in production.[152] Both Robertson and Gaultier corresponded on Hitchcock's behalf, but Robertson's letters are more frequently about business she's conducting alongside Hitchcock than missives typed for the director.[153]

On the set, Robertson acted as both filter and proxy for Hitchcock, typically arriving at the crew's call time to "fuss around and make sure everything was ready that Hitch wanted and he'd turned [over] to me and I would have checked everything out for him."[154] As the day went on, she would dispense with lower-level tasks that she deemed beneath the director's notice: "They would be lining up so many things to do. They came from all sides. Suddenly you're in the middle of something and then you get a call from the labs. . . . The cameraman hasn't got his lens. . . . Actors don't know their lines. We've got some cut film to run for Hitch. There was never a time when you'd sit back in your chair and say, 'Oh, well, *that's* done for the day.'"[155] She also rehearsed actors for the director, deferring to him on any matters that required special attention, such as a particular line reading. At other times, she functioned less as Hitchcock's assistant, and more as a member of the team responsible for managing the work of production. She filtered out "small problems [Hitchcock] wouldn't deal with," but did so as one member of this small group, among whom work was distributed according to specialty ("You had calls for the production manager, for the location manager, for the assistant director. Everyone did their part, however small").[156] Robertson's territory generally included issues related to casting, actors, scripted material, and other aspects of the film's planning process.

Unique among the women described in this chapter, when Robertson took on a problem on behalf of Hitchcock, she tended to take on all duties associated with it, and he left her to it. It was Hitch who established this autonomous style early in their work together. When, for tax reasons, he was required to leave England after production wrapped on *Stage Fright* (1950), he left Peggy in charge of post-production on the film. The decision, and the actions he took to enforce it, demonstrated both his faith in her ability to manage in his stead as well as his understanding that this decision did not reflect typical treatment by male employers of female employees. As Robertson recounted, Hitchcock used his persona—fascinating to his employees—to subtly convey a message he wanted carried throughout production:

> So the studio was furious, the sound man and . . . "A woman in charge? What's the matter with the old boy?" I mean they *hated* it and *loathed* me. Hitch used to send cablegrams—he was back in Hollywood now—which were very famous, because they were always very long, and he sent me a ten-page cablegram. "Dear Peggy. Alma and I . . ." This is in a cable, mind you. In those days, one just sent "Happy Birthday." "Should I put 'love' or not?" That sort of attitude. So, "Dear Peggy, Alma and I were having dinner last night, Tuesday January the fifteenth," or whatever it was, "and Alma said to me . . . 'Who is going to finish STAGE FRIGHT for you if you're going to be over here?' So I thought for a minute . . ." and it goes on like this in the cablegram, and I'm reading this, and the end of the cable was, of course "I said to her, 'I think Peggy would be a good person to finish the picture.'" Alma said, 'That's a wonderful idea, Hitch.'" . . . Well, I went around the studio showing everyone this cablegram, and everyone burst into laughter. We thought this was the funniest thing. . . . I dined off that story for many days. Well, I realized much later on, that he'd done it on purpose. He *knew* that this was funny telegram, cablegram, he *knew* I would show it around. . . . He *knew* that it wasn't a popular move putting me in charge. This way, it would be a bond. I said to him once, "You really did that cablegram on purpose, didn't you?" He just smiled. He knew. He knew I was onto it.[157]

In this instance, Robertson is granted substantial managerial power on Hitchcock's behalf. But she showed real creative agency between productions when she drove the process of finding new material for Hitchcock to direct. Though, as Robertson said, "that was one of my jobs, looking for the next project," it was still Hitchcock's job to decide what to make from the properties she selected. Her creative impact came in the form of presenting the best few candidates to the director.

The offering of choices, as Robertson did during Hitchcock's development process, remains a key characteristic of the work of today's assistant—a title that describes all clerical/administrative support roles in the contemporary media industry, where the distinctions among secretary, assistant, and executive assistant have been flattened or eliminated. In selecting a handful of what they deem the best available options, the assistant facilitates the solution to a problem by delimiting its outcome. This task is usually a matter of assembling a list or surveying a field of possible solutions to a problem, creative or otherwise, weeding out the less suitable possibilities, and presenting the decision maker with the best few from which to select. This practice, long used to solve practical problems for an employer (for example, assembling carpet swatches in the five best shades of blue) has become more important in the development stage of production since the end of the contract system, as the process of finding and financing material has become more independent and individualistic, based on the employer. In Robertson's case, narrowing the field of options for Hitchcock's next project was largely creative work, requiring both creative sensibilities and good taste; however, the work was still done on behalf of her employer, with his sensibilities in mind, requiring her to hand off the decision rather than make it herself.

Hitchcock, who had more leeway in his choice of projects even in the studio days, relied on Robertson for much of his development process. Peggy read incoming material from agents and studios. "He *expected* me to read, and I used to read all the coverage that came in from agents," as well as "maybe ten a week from Paramount," or at MGM during Hitchcock's time there. In reading these books, scripts, or coverage on Hitchcock's behalf, Robertson was looking for "a feeling" that the material could lend itself to Hitchcock's style and tone.[158] With the exception of stories Hitchcock discovered on his own, Robertson effectively delimited his choices by eliminating unfavorable candidates and only putting scripts in front of him that she thought should be made. And, because Hitch trusted Peggy, she might also influence his final decision. The same was true of the casting process, where she made her own lists of casting suggestions for Hitchcock, just as the casting director would do.

Robertson also affected the creative process through her choice of films to screen for Hitchcock as he was preparing to go back into production.[159] In her movie selections, Robertson followed her own tastes: "I'd just make out a list myself of all the films I wanted to see, and send it to the woman who ordered them in the cutting rooms."[160] When asked what films Hitchcock would be watching if he were still alive—in 1995, at the time Robertson was giving her oral history—she cited "*Muriel's Wedding,* and all the new Australian ones which are out. He'd have seen all those because I'd want to see them."[161] If we understand Hitchcock's visual style and vocabulary as enhanced by the films he watched, then it becomes clear that Robertson—in her capacity as movie selector—shaped his creative process.

Robertson's impact is quite clear in certain films. She acknowledged her own creative contribution to one particular production, asserting, "I found *Marnie,*" recounting how she read the book on a Friday night and called Hitchcock the next morning, insisting that he read it.[162] She also suggested Sean Connery for the male lead—the role in which he was later cast.[163] Thus, for a single film, Robertson might have found the source material, chosen the lead actor, and even influenced decisions regarding film style and technique based on screened films.

Measuring the extent of Robertson's role is not meant to suggest that Hitchcock—well known for meticulous storyboarding and a unique visual style—was not chief author of his films, but rather, that if filmmaking is a process involving multiple participants, an assistant or secretary's participation should be given some weight alongside that of supporting crew and technicians, such as cameramen and electricians. In setting the parameters for Hitchcock's choices through her own taste and judgment, Robertson functioned much like a contemporary assistant or creative executive to a producer during the development process. However, though later credited by some with having been Hitchcock's producer, Robertson is still invariably referred to as Hitchcock's assistant, if at all.[164]

As with the other notable secretaries and assistants discussed in this chapter, Peggy Robertson's level of creative agency was possible because her employer sanctioned it. Though she received fair treatment from Hitchcock, sadly, he would have been within his rights to treat her otherwise. This alternative was clearly the case even after the end of the studio era, as evidenced by accounts of Hitchcock's behavior toward other female employees, most notably Tippi Hedren on *The Birds* (1963). Though these accounts vary as to the severity of his harassing, sexually abusive behavior, most agree that, at some level, this conduct did occur. Hedren recalled the same code of silence around such behavior, still in place in 1963:

I had not talked about this issue with Alfred Hitchcock to anyone. Because all those years ago, it was still the studio kind of situation.

Studios were the power. And I was at the end of that, and there was absolutely nothing I could do legally whatsoever. There were no laws about this kind of a situation.[165]

Hedren claims that when she asked to be let out of her contract after finishing *Marnie*, Hitchcock essentially ruined her career. She no longer had to work with him, but under the terms of the contract, he was able to accept or reject offers for her, and he rejected nearly all proffered roles until the contract period ended several years later. Though she and Robertson remained friends after Hedren's association with Hitchcock ended, the actress later said that when she appealed to Robertson, she "did nothing, beyond trying to placate me and help get the movies finished."[166] Robertson's creative and professional fate was in Hitchcock's hands as much as Hedren's was. After his death, despite twenty-two years of service to the director, Robertson was surprised to discover that he had made no provision for her in his will.[167]

Conclusion

The daily activities of each of the secretaries discussed here differed enough one to the next that it might be possible to dismiss their work—from Valeria Beletti's participation in recutting *Stella Dallas* to Peggy Robertson's supervision of postproduction on *Under Capricorn*—as the product of their employers' individualism, resulting naturally from a combination of the practices of a movie maker (who happened to be male) and the skills of a secretary or assistant (who happened to be female).

However, when treated as different iterations of the same creative service, the whole of the secretary's labor becomes more visible, as all of the tiny unimportant "details" dismissed as irrelevant to creative production (the tactfully delivered message, the confidential personal errand, the pleasant voice at the end of the line) add up as contributions to (or often, subtractions around) the creative process. Reframed in this way, these different stories of secretarial work are revealed as direct products of the same logic around women's labor: organized by the studios' practice of sex segregation and further shaped and enforced through their gender-normative culture. Much of women's labor at studios functioned in the same way, as a more general form of creative service for creative labor studio-wide, removing as it did all obstacles from creative workers' paths so that they could dedicate themselves to the production process.

In reality, this work system did not evolve to serve great art or provide authors with their truest means of expression, but to harvest the most work from the cheapest workers. Women's labor, a bargain to begin

with, only accrued value over time, as women expanded to meet subtle but implicit expectations held of their gender. Because they had no
other avenues to more satisfying work, they often carried out impact-
ful tasks without being asked directly to do so, and without asking for
recognition in return. Incidents of significant creative intervention
(for example, Koverman with Garland, Robertson with *Marnie*) can
be found in the accounts of every secretary whose career is examined
here. Unfortunately, such detailed accounts of specific individuals are
rare; thus the spirit of these stories must stand in for the impact made in
other offices, by other women.

Samuel Marx, one of the few men to devote more than a line of his biography to female workers' role in his own success, said of the MGM secretaries
he oversaw as story editor: "Young, personable and ambitious, some of them
became writers. Many attached themselves permanently to the men they
were temporarily assigned to—as collaborators, housekeepers, mistresses
and other things."[168] A *New York Times* article about screenwriting also noted
the impact of the secretary's labor on the process, explaining, "Almost anyone
can write, but few can be studio secretaries. The one you have has worked for
three or four hundred dramatists, knows all about screen technique, camera
angles, exits, suspense, climax, the clinch and fade out to full orchestra music.
But since she isn't known as a writer she remains forever a secretary—to your
good luck."[169] These accounts recognize the unevenness of the exchange that
many secretaries made in order to participate in creative work, relinquishing
all claim to appropriate credit and compensation for creative contributions
for the privilege of being allowed to make them.

Secretaries' wages reflected the unevenness of the trade as well.
Though hers was a good wage relative to other secretarial positions,
Valeria Beletti points out that in comparison to the other staff assigned
to the production of *The Winning of Barbara Worth*, her salary doesn't
hold up: "Frances Marion will get $10,000 for adapting the story,
[Henry] King will get $75,000 for directing it, assistant director gets $500
weekly, art director $400 weekly, technical director $300 weekly, Ronald
Coleman $1750 weekly, [Vilma] Banky $1000 weekly. . . . Of course you
mustn't forget my $40 per—it makes me sick when I look at our payroll
and then look at my salary."[170] A few years after her stint for the demanding Goldwyn, when she was forced to choose between a larger salary on
an executive desk at MGM and a smaller salary working in the script
department, taking dictation from its writers, Beletti did not hesitate to
choose the latter, explaining:

> If I get in with MGM and with Frances Marion on the lot, I will
> have a fine opportunity for working myself up. You see I will come

in contact with all the writers and will have a chance of studying their technique when working for them. MGM start their girls at $25 per week . . . and at the end of the year, a girl (if ambitious and intelligent) is usually assigned to some writer as secretary and the salary ranges from $35.00 to $50.00. While this is not as much as I would get as a secretary to an executive, I think it will be better because I won't have to devote all my time to one man and be so exhausted at the end of the day that I don't want to do anything but rest.[171]

For Beletti, the high-impact work of clearing creative/executive space for Goldwyn was not worth the few dollars extra it paid over a post with more creative rewards. Better to hope for creative involvement in service of a writer and in the vicinity of Frances Marion, whose offer of mentorship was still the best guarantee a woman had of training and opportunity in the 1920s.

It is hard to view these realities of secretarial work in either a wholly positive or wholly negative light. Studio secretaries were able to participate in otherwise off-limits creative sectors by offering creative service rooted in feminized labor and disguised through feminine performance. However, the limits placed on their impact and the lack of credit they received betrays just how much more of their potential contribution was likely suppressed because it didn't fit the studios' interests in cheap labor, or their image of women as men's little helpers.

More important than characterizing the studio secretary's history as a net win or loss, is recognizing that it isn't (just) history. Generations of female production workers inherited the studio secretary's fate. Production specialties that emerged as women's work during and after the studio era followed many of the same unwritten rules of feminized creative service as the secretarial profession. On film sets and in production, female workers were routed into fields where they again functioned as the filterers and emotional managers of the creative process. As women broke into masculinized fields, the fields themselves assumed many of the service characteristics attached to other feminized work. These gender ascriptions towed the line between men's work (at the creative/managerial center) and women's work (at the noncreative/feminized margins), so that a woman might ascend to the level of a casting executive and still be nowhere near the level of director.

The next chapter unpacks this process of feminization in women's production specialties as it developed in the studio and post-studio era. Tracing women's professions back to the provenance of their feminization helps link them to their counterparts in the present, many of which remain female dominated and all of which preserve restricting feminized characteristics.

5

Studio Girls

Women's Professions
in Media Production

Then comes the continuity writer, the architect,
who must know and apply all the laws of screen
drama in translating this dream into visual action....
Why do it if it is so hard? Because it is so hard!

—Jeanie MacPherson, continuity writer
and screenwriter, 1922

In a sense, all traditional women's labor sectors at studios (for example, domestic service, light manufacturing), as well as the film-specific professions that arose within them (for example, inking and painting) served a global function of creative service similar to that of the secretaries in the previous chapter, with women's labor functioning, in essence, to absorb routine tasks and unwanted emotion around men's creative process. This gender dynamic was particularly pronounced in the clerical divisions that employed so many female workers in the studio era, as well as in the many film-specific women's professions that descended from them between roughly the 1920s and the 1970s. Some emerging production sectors feminized precisely because of their resemblance to secretarial work; others became accessible to women because their processes took place on paper and required a nearby staff of predominantly female clerical underlings to carry them out. In some cases, gender-neutral or masculinized jobs carried

Fig. 24. MGM research department, circa 1939. (Courtesy of the Academy of Motion Picture Arts and Sciences.)

latent aspects of creative service, which contributed to an impression that they were good jobs for women. Still other jobs shifted toward feminization after the end of the contract system, when formerly executive roles became freelance crew positions of lower status within the production hierarchy.

When female workers took over jobs that had previously been done by men, the professions shifted to include greater components of emotional performance and detail work, compelling their own forms of creative service from female professionals. Women actively reshaped these jobs by playing to their strengths (perceived and actual) in order to succeed and to prove themselves assets to production. In many cases, these strategies led to greater responsibility and amplified roles in the creative process. But, even when women increased the creative capital of their professions, their work remained underpaid, underestimated, and undercredited in comparison to men's. This process of feminization greatly affected the media industry's own evolution in terms of not only gender integration, but also the way workers identify within professional groups, their creative practices, and the products that result from them.

This chapter further explicates the feminization process and its downstream effects on media production history by tracing the origins of the contemporary women's professions invoked at the start of this book. Firsthand accounts from workers offer rich material about the process and effects of this trajectory.[1] The production specialties that feminized earliest are those most closely linked to feminized clerical labor through their women's work components and the forms of creative service they elicited. These specialties include the various production sectors that today are more commonly situated under the heading of development, as well as script supervision and publicity. The origins and gendered aspects of other female-dominated professions are less apparent. One such profession—editing—involved little clerical labor or paper planning, nor could it be said to have been truly female-dominated or feminized at any time in film history. However, it is historically understood (by industry and the academy alike) as female-friendly, and represents a kind of boundary case because of characteristics it shares with other women's jobs. Linking editing to more unambiguously feminized jobs reveals how the industrial and cultural logic that guided feminization spilled over into nearby roles, limning an outer limit for this study's provisional definition of women's work in media production. Connecting editing to feminized fields through this logic of feminization also helps explain a second boundary case: casting. Casting presents an unusual example because, though casting departments had been more open to women than others in the studio era, the role of casting director feminized relatively late in the industry's history (in the 1960s and early 1970s), and today holds relatively high creative and managerial status. This field's arc from being masculinized to heavily female dominated reveals at work the same logic that guided feminization decades earlier, and that causes some production sectors to continue to be associated with women today.

Readers, Weeders, and D-Girls: Creative Service in Studio Era and Contemporary Media Development Roles

Increasingly in the 1920s, women came to screenwriting jobs not from writing careers outside the film industry (though many had them), but through the lower ranks of studios' literary or intellectual departments—those related to scripted material—and secretarial roles. Like Dorothy Arzner, Joan Harrison, and many of the other female writers previously mentioned, Charlotte Miller was promoted to screenwriting from a secretarial position. Though Marguerite Roberts had been a reporter before she came to Fox in 1926, she, too, did secretarial service on the desks of various studio executives, and then worked as a reader before ascending to writer.[2] Isobel Lennart followed a similar progression from stenographer to script girl to reader, and through MGM's junior writer program, before advancing to screenwriter.[3] These promotions, on their surface, evince studio management's gender-blind, even feminist, attitude toward screenwriting. However, placing these accounts of women's screenwriting in the context of women's work in nearby reading, research, and story departments reveals that the female screenwriter's ascension was granted by the same inductive reasoning that contrived to keep women out of most studio-era production jobs by confining them to a few feminized associations.

The early association of women with the scenario-related departments described in chapter 1, such as reading, research, story, and script (a.k.a. stenographic), occurred via the same logic as their association with screenwriting: feminine intuition, morality, and a flair for sentimentalism and drama were assets in the invention of stories—particularly those geared toward female audiences. It was often women themselves who made this connection in the early years of film. Through writings on women's place in production, they attempted to carve out a niche for themselves while reassuring male colleagues that the advantage they offered was complementary, not competitive.[4] Alice Guy Blaché declared women "an authority on emotions" because of centuries of indulging them under the protection of men.[5] June Mathis claimed that female screenwriters complemented male directors just as wives complemented husbands, delivering "the sleeping soul of drama" in a pile of typed scenes for the director to animate into physical form, ensuring that the "human quality that makes a motion picture commercial" would not be lost in all of his technical prowess.[6] She also insisted that women were more suited to the "careful, fine detail work of scenario construction."[7] Florence Osbourne cited scientific studies of children and storytelling that concluded that although "girl children were not so logical in their plots," their stories showed "more interesting detail and more sentiment and more emotion."[8]

Though offered in support of female screenwriters, these rationales endorsed gender binaries that undermined women's prospects elsewhere. Implicit in discussions of women's aptitude for storytelling was the belief that the female screenwriter's value issued primarily from gender rather than from her individual talent for plot or story. Such gender-based value typically did not travel beyond feminized or female-friendly sectors at studios, limiting women to a narrow range of fields. Declarations of women's natural aptitude for writing effectively doubled as rationalizations for why women could not or should not succeed beyond that realm. Jane Murfin described the impact of this double bind: although studio managers believed that "the feminine slant" was necessary in writing, women's gender was her chief handicap in accessing more visible financial or managerial leadership roles, such as producing and directing, and thus in the larger industry, because "the more jobs she has filled, the more she knows of the business as a whole, the better chance she has of making a real place for herself."[9] Frances Marion believed that women succeeded in writing less because of innate, feminine suitability than because they could do the work from behind the scenes, drawing little attention to themselves.[10] In many respects, female screenwriters at studios, though members of the top tier of the scripting hierarchy, had as much in common with the women in the departments below them as with the male colleagues who populated the upper-level creative fields of directing and producing. For, like screenwriting, work in research, reading, and story departments all centered around typing, took place on paper, behind the scenes, and away from set, and made use of women's "natural" talent for detail and judging emotions.

Like stenography in script departments, or secretarial labor in the service of writers, producers, and executives, film research at studios required heavy components of clerical labor, a fact reflected by researchers' inclusion in some studios' office employee guilds and descriptions of the job as requiring typing skills.[11] And, because of the educational focus, clerical, and service characteristics and the feminized associations of library work in general, women tended dominate positions in research departments at studios as well.[12] A 1930 feature in the *Talking Screen* entitled "Their Business Is Looking Up" profiled research at Fox, MGM, and Paramount and described only women in the research roles.[13] Trade articles and vocational manuals from the 1930s also characterize research departments as largely or even exclusively in the hands of women and only female researchers are visible in the studios' 1920s and 1930s promotional tour films (see appendix).[14] In another of his candid statements about why women do certain work, director/producer Mervyn LeRoy explains that

> research is one department where mature women, preferably unmarried, are employed. There is an enormous amount of work to be turned out, and great responsibility. . . . Unmarried women are preferred because they are less independent about their jobs. If a married woman stays out for a day, she still eats. This may be a rather cool appraisal of the situation, but with a limited staff, and the production going at top speed, the girl who is dependable, always there, not worrying about her husband's dinner or her child's cold, makes the best researcher.[15]

At many studios, women also led research departments. Elizabeth McGaffey founded the department at Paramount (then Lasky) in the late teens and oversaw it for many years thereafter.[16] Natalie Bucknall, described by Samuel Marx as "a solidly built Russian woman reliably reported to have ridden with the Cossacks in World War I," managed the research department at MGM until the 1950s, and can be seen in a photograph of the department from 1938 presiding over four female and three male workers.[17] When LeRoy wrote his how-to-get-into-pictures book in 1953, Dorothy Luke had taken over the MGM research department. He explained women's leadership as typical because "women are much better than detail work than men, who are usually bored by it. Just as there are few male librarians, there are extremely few men researchers in the studios. For one thing, men would not be content with the established salary, which is fairly high [$82–$108/week] for a woman, but is not particularly high for a man."[18] Though Carl Milliken headed Warner research in the 1940s, his staff was predominantly female (eight out of a total of twelve workers).

Milliken's 1940 *Warner Club News* cover story on the studio's research department speaks to the work's clerical and organizational aspects, despite the fact that many who worked in research had advanced degrees. Milliken states that the department's business is not in knowing everything, but "merely in knowing where to look for the desired information" in a library of seven thousand books and hundreds of volumes of magazines.[19] Descriptions of work at other studios cited multiple clerical and service characteristics as well, including gathering materials, keeping ahead by "storing up data and preparing against possible emergencies, in addition to supplying facts for current pictures," working "hand-in-hand with costume and art departments" and "writers, wardrobe and make-up personnel on every picture which involved historical detail," answering continuous telephone inquiries, and poring "through voluminous masses of matter, keyed by the files and indexes, to obtain various facts."[20] Researchers are also often described in terms of the errors they subtract from scripts (per LeRoy, "If a scene shows a star reading a book that has not been written at the time of the picture . . . it's the researcher who takes the rap").[21]

Taken together, these job duties—performed in service of a film's design team—begin to resemble the creative service offered by the secretary to a movie maker. Like the secretary, the researcher's job was to filter out all information except that most directly connected to the creative process, such as narrowing a field of photographs or drawings down to a few choice examples, which could then be offered to an art director or costume designer. Researchers affected the creative process subtly—often through subtraction—shaping a production's aesthetic through the range of plates, photographs, and drawings they offered (and the larger number discarded as unsuitable)—as well as by their answers to queries and by checking scripts for "mistakes and errors—anything out of period."[22] Research also shaped the direction of projects in the scripting phase, as in the case of Howard Koch's script for *The Sea Hawk*, which changed directions based on historical background information assembled by Warner researchers.[23]

Fig. 25. June 1940 *Warner Club News* cover image, which accompanied the "Information Please!" feature story on the studio's research department. (Courtesy of the Academy of Motion Picture Arts and Sciences.)

Occasionally, researchers made more direct impacts on a picture that could be identified as theirs. Natalie Bucknall persuaded Ronald Coleman to "shave off his famed trademark mustache—or run the risk of its being an anachronism" for his role in *A Tale of Two Cities*.[24] However, in most cases, researchers supported creative work from their libraries, shaping artists' and writers' creative parameters (for example, in delineating which types of architecture could or could not be used to represent a specific period), themselves having little direct participation in final creative decision making.

Readers also shaped solutions to creative problems—deciding which literary properties to make into films by weeding out materials from a field of candidates, and recommending those that remained in "coverage"—a typed plot synopsis and comments assessing the property as a prospective film. Kate Corbaley of the MGM story department described dramatic sensibilities as foremost among the many competencies necessary for success as a reader, explaining, "Well knit drama does not happen; it is evolved from adherence to certain laws of building." In addition to an innate sense of whether a story worked, these practical business concerns had to be factored in when considering material. To be successful, she continued, a reader "must be capable of deciding whether or not a story will fit the needs of the individual studio in which that reader is employed" by weighing its suitability for the company's actors, its cost, the studio's other recent films, and the public demand. To this considerable list of the reader's qualifications, she added awareness of other material on the market to guard against plagiarism, as well as knowledge of continuity to prevent the reader's judgment from being swayed by "the beautiful phrases, by the colorful and sympathetic language in which a story is often clothed" in order to correctly estimate its screen value.[25]

Given Corbaley's description of the reader's skills, as well as the importance of synopses and comments in the acquisition process as materials made their way up the chain of command to decision makers, one might expect readers to have elevated status within scenario divisions' hierarchy. And, initially, studios did seem to value their readers. In 1922, Corbaley reported that reader salaries were increasing as studios sought "men and women with the qualifications a reader should have, for present day producers realize the paramount importance of the story."[26] Yet this was hardly the case in the fully developed studio system of the 1930s and 1940s, nor is it so in Hollywood today, where, with few exceptions, the reader has no place at all in the hierarchy and in most cases freelances from home, with contact only via e-mail. Practically since its inception, the job of reader—variously referred to as "screen analyst" or "story analyst"—has been seen as drudge work, even though it was also a stepping stone to screenwriter or story editor roles. Many screenwriters—including Budd Schulberg, Philip Dunne, Lillian Hellman, and Lester Cole—began their film careers as readers.[27] Hellman and Cole worked together in "a large office with a dozen or more other readers," tasked with covering one story per day, though Hellman quickly lost her job by inciting her fellow readers to demand higher wages.[28]

And, at its most basic, the work of story editors like Kay Brown, who first brought the manuscript for *Gone with the Wind* to David Selznick's attention, was similar to that of readers, but on a higher level.[29]

However, until the transition to writer or editor occurred, readers were closer to the anonymous movie workers Leo Rosten characterized as swarming over "the offices wherever pictures are fabricated" than to movie makers doing exclusively creative work.[30] Samuel Marx dispelled misconceptions of MGM's reading department (which, from the outside, resembled a library reading room) as "a hideaway for those who wanted to nurture their own writing efforts," instead characterizing it as "a hotbed of frustration" because of both readers' "cerebral indigestion caused by a diet of materials they knew they could cook up better" and the "absurdly low" pay of the work, which studios could offer because it was nonetheless sought after.[31] While the readers' bitterness over low pay is partly explained by this description of frustrated writing ambitions, another likely cause was that—despite the heavy concentration of highly educated, middle-class literary figures like Hellman and Cole in its ranks—the studio reading department was bordered on nearly all sides by feminized sectors and, in many studios, was female dominated and led by women.

Though Marx referred to readers as intellectual staff—in keeping with their supposed educational backgrounds and writing and editorial ambitions—the work was also linked to clerical sectors, with its heavy typing and organizational duties (for example, logging submissions, summarizing contents via coverage, corresponding with writers, maintaining files of material), as well as because so much of the reader's time was spent filtering out unsuitable material (the vast majority) in much the same manner as a clerk or secretary winnowing routine tasks away from an employer. Marx wrote of the importance of this process, which allowed him to pass select material upward to Irving Thalberg and his supervisor-producers: "Our coverage averaged four hundred possibilities a week—books, plays, news items, magazines, original ideas—all flooding in from God's huge writing staff, slaving away twenty-four hours a day from all around the world. The story department checked twenty thousand submissions a year and all of them crossed my desk."[32] Despite readers' pivotal role in discovering the material that would ultimately become MGM films, those films would be credited to their writers, producers, directors—and to Thalberg. Though story editors like Marx also helped narrow the field, the ratio of weeds to flowers was much greater for the readers—the studio's primary filters of material—and their participation in higher-level decision making or creative processes was nil. In this sense, the readers' work might also be viewed as creative service, in that it cleared the way for the intellectual sector's creative solution by taking on low-level creative tasks to free executives up for higher-level work.

Like the clerical, library, and educational fields with which it shares common characteristics, reading in the early twentieth century was female friendly and in some cases, female dominated. Long before Marx oversaw MGM's reading department (itself supervised by Dorothy Pratt, "a lady with an incredible memory for every story in its huge files"), the field was considered one of a few wide open to women.[33] When an anonymous scenario editor singled out readers for scorn in 1919, he described reading departments (along with clerical staffs that supported editorial departments in general) as characteristically feminized and female led:

> When manuscripts come in they are handed over to the reading department. This is a room where half a dozen women at an average salary of ten dollars a week, without the competence of a stenographer or salesgirl, sit all day making first choice of the material the editor is to see. . . . Whether there is some vague notion on the company's part that these young ladies represent the typical motion-picture going public mind or not, I cannot say; but I believe not. It is simply that they seem to form a necessary machine for weeding. When one of these girls finds a story which she considers would make a good picture, all she has to do is fill out a form.[34]

The low cost and high availability of feminized labor in the late teens made it easy for male superiors to view readers as yet another feminine paper-processing system—a machine in which ever-multiplying submissions could be dumped on one end so that select stories might emerge from the other. One vocational manual characterized reading departments as "often filled by women."[35] Similarly, departmental columns in studio newsletters indicate that reading staffs—though not made up entirely of women during the classical Hollywood era—were similarly female friendly and/or female led.[36] By the 1950s, when LeRoy wrote about readers in his vocational guide, he used male pronouns in general statements about the typical reader's "good analytical mind for stories" and "command of fluid, pure English," but the specific readers he cites as real-world examples are women. He gives a hint to the reasoning behind the mixed-gender makeup of studio-era reading departments when he explains that readers have to be a cross-section of the population, saying "An even number of men and women is wanted," but adding that "as a rule, a man is better qualified" for "certain highly technical books on aviation, or on naval maneuvers."[37] This statement perhaps explains why men were employed in a department with considerable clerical requirements ("Your letter and your sample synopsis should be impeccable") in higher numbers than in nearby female-dominated departments, at a starting salary that LeRoy reports at $74.60 per week—only marginally higher than the secretary's $56.60–$63.00 per week (depending on experience) and lower than the researcher's $82 per week.[38] Marx reported that readers were paid an "absurdly low" $50 per week at the same studio in the 1930s.[39]

In truth, the reader's pay was only low compared to nonfeminized work. For female secretaries providing creative service, whose salaries ranged from $35 to $65 per week, $50 was the going rate. As for nonmonetary compensation (for example, credit for creative contributions), the reader's was almost as low as the secretary's, as was their status in the hierarchy. That the lack of respect afforded the reader had at least something to do with gender seems obvious in early, male-authored accounts of "girl readers." Despite the creative sensibilities, taste, and analytical skills required for their work, readers were—and still are—regarded as insignificant within the studio hierarchy, just like stenographers in script departments, secretaries to writers or producers, researchers, or any related women's clerical position.

Though she seemingly remained behind the scenes and out of visible leadership by her own choice, Kate Corbaley, who all but ran the MGM story department in the 1920s and 1930s, embodied the poignant contradiction between the value of women's labor in literary departments and the way studios acknowledged that labor. Discovered by Hunt Stromberg while working as a librarian, Corbaley accompanied the producer first to Ince Studio and then to MGM, where she was still working when Marx "inherited" her as his assistant.[40] Marx's writings infer that she might have had his job as story editor if she had wanted it, but she remained below him in the hierarchy by her own choice because she was singlehandedly raising four daughters. He suspected Corbaley was loath to leave her children even for a day, much less to travel to New York or Europe, as he did for his role. He also recognized that she was his superior in the esteem of nearly every producer and executive in the story meetings they conducted; as Frances Marion remembered, "Supervisors and minor officials gathered in Kate Corbaley's office to hear her weekly analyses of all the published material submitted to the studio."[41] Marx had no doubts about who, between the two of them, was running the meetings:

> Corbaley was the star. She sat at the head of the table. Thalberg was a supporting player, down at the other end, a rapt listener. Story telling is surely one of the world's oldest arts and many Scheherazades are noted through history, but there was a special finesse to Kate Corbaley. She never forgot a detail, never needed to retrace her steps, failings common to less talented yarn spinners. She injected color, clarity, and characterization that might well have won the envy of the original author. Because she literally thought in pictures, she could embroider a plot into a movie with elements its creator might have

overlooked. Our listeners knew this. They were absorbed by the plot turns she offered them. . . . Bernie Hyman, who had accepted a book Kate told, lamented that after reading the material, he found it far from what he had heard. Thalberg shrugged and said, "shoot the story Kate told!"[42]

As Thalberg's Scheherazade, Corbaley performed script coverage, guiding studio executives who "were too busy with their own problems to concentrate on the written word," toward the stories they should buy.[43] While never overlooking a detail herself, she filtered the material for those above her through her omissions and additions, directing their attention via her delivery to those story elements she deemed important. Louis B. Mayer refused to be read plot summaries by anyone but Corbaley. For Marx, Kate's way with material "pumped the lifeblood of the story department," a fact not lost on Mayer, who, at Corbaley's funeral in 1938, leaned in to Marx and said, "I would rather have lost any star than this woman."[44] And yet, her creative contribution to the finished product is essentially invisible to us now. She is little known in film history, much like many of the women whose labor—clerical and creative—supported her department.

The multidepartment development sector still exists in various forms today, at film studios, networks, production companies, major talent and literary agencies, and anywhere else that film and television development processes take place. And though screenwriting itself is no longer done in house, a mixture of senior-level executives and producers, junior-level creative executives, and their underlings still carry out the rest of the process of searching for material, developing it into a fully realized screenplay, and securing production financing. These underlings, including story editors (who often double as assistants), producer and executive assistants, development assistants, story analysts (a.k.a. readers), and interns, sort good material from bad. Among them, these workers accept submitted materials, assign them for coverage (which still includes a plot synopsis and comments in roughly four pages), type and organize records of submitted material and the communication around it, and circulate that information, along with various drafts of purchased scripts in development to the executives. Thus, the modern-day assistant in development descends not only from the studio secretary, but also the reader, the researcher, and story department workers like Corbaley.

Though some networks, studios, and agencies employ a staff of readers housed onsite, for the most part, freelance readers conduct story analysis, working remotely and contracted on a script-by-script basis.[45] Research libraries have been downsized or dismantled and sold off by many studios and have never existed at most of the smaller production companies and "major independents" where so much development takes place. Development-stage research often falls to either the script's writer (who may pay a researcher out of his or her salary or receive additional funding for this step) or assistants and interns. The gender breakdown of present-day development workers is impossible to calculate given the ad hoc arrangements that spring up and dissolve under the current flexibility of the post-Fordist, post-digital industry. However, the derogatory slang term "d-girl" or "development girl"—which only fell out of use in the past ten years—bespeaks the deeply feminized origins of the various lower-level development jobs that, until recently, were carried out by a predominantly female group of workers, under a predominantly male group of manager-creatives, just as they had been in the studio period.[46]

Whether male or female, assistants still carry out the bulk of the labor required for development, and do so for low pay, little credit, and the faint promise of promotion.[47] Yet women's professional realization in these lower-level development fields has been far more fruitful than in continual (frustrated) efforts to be hired in equal numbers to men as screenwriters— a hardly surprising situation given that for women at studios, feminized, script-related departments offered some of the only paths to screenwriting. In the wake of the contract era, with the separation of screenwriting from its low-level sister fields because of the shift to freelance, aspiring women writers had a far harder time selling their talents on spec than being hired to assist, read, or research. Whether they were or are fairly compensated for development work is another question; low wages remain an industry standard. Most modern story editors outside of those at major studios work for an assistant's salary—$30–35,000/year—and often assist executives on top of their story-related responsibilities. Increasingly, the other substantial bottom tier of script-related labor is delegated to interns who do it for free, or freelancers who are underpaid for it with the implicit understanding that it is relatively creative labor and therefore mythologized as a reward unto itself.[48]

Various chick-lit series set in the media industries celebrate the modern, typically female assistant's "development hell."[49] The stories often modernize the studio-era office Cinderella trope by launching the heroine into a successful career at the end of the story, but only while simultaneously launching her into a romantic relationship, usually with the boss or some other authority figure from the office. The price of these books' happy endings is a particular kind of torment on the part of the female assistant or secretary over the course of the narrative at the hands of a high-status media maker. A comparison of 1938's *I Lost My Girlish Laughter*, discussed in chapter 3, and 2004's *The Second Assistant* reveals almost identical plots. Rather than being eradicated over the intervening decades, the culture of abuse and the mythology of suffering abides around women's work in service of the development process. This book's epilogue revisits these development workers, surveying contemporary assistant work and its inherited feminized traits, including its trope of suffering.

Script Girls and Cutter Girls: Women in the Space of Creative Play

Beaulah Marie Dix described the role of script clerk in its earliest iteration, saying that in the improvised production systems of the early film industry, "anybody not doing anything else wrote down the director's notes on the script."[50] In some cases, it fell to the cinematographer to watch actor entrances and exits and keep "accurate account of every motion made," but the practice was informal and idiosyncratic.[51] Generally, Janet Staiger explains, continuity and verisimilitude were the responsibility of the director, cameraman, and staff, in much the same catch-as-catch-can arrangement Dix observed. This procedure persisted until the mid- to late 1910s, when "demand for accuracy increased the need for paper" as productions developed more complex lighting techniques for more intricate scenes shot from multiple angles.[52] Tracking continuity between setups and in shots taken out of sequence became increasingly time-consuming, intricate work for which casual approaches would no longer suffice.

Between roughly 1917 and 1920, the specific production role of script clerk, more commonly known to as the script girl, emerged, as companies added devoted workers to their staffs to track continuity on paper.[53] Sometimes referred to as an "assistant," "script assistant," or "continuity clerk," this member of the production team took notes during each shot for reference in later filming and during the editing process and was responsible for minding properties, costumes, and noting any deviation from the script as written.[54] In 1918, *Photoplay* reported the employment of continuity minders as common practice, stating, "Nowadays in the big companies the director has an assistant, usually a stenographer, who keeps track of all the details of every scene so that mistakes are almost impossible."[55] Though the job may have fallen to male crew members prior to its separation from other production tasks in the 1910s' process of specialization and standardization, once clerical tasks were consolidated into one dedicated position, the record-keeping role would have had little appeal to men in camera or other technical departments.

At the same time, women seemed like natural candidates for work involving note-taking, details, and stenography. Even in the early days, Staiger reports, the script clerk was "usually a woman." June Mathis wrote that the woman script clerk emerged shortly after the woman scenario writer, and for much the same reason. Just as women succeeded "in the careful, fine detail work of scenario writing," on set, they could also "watch the small details better than men."[56] But even as the job evolved, becoming more clearly coded as clerical (involving typewriters and stopwatches), male script clerks were used in certain productions. Barbara Hall cites accounts of male script clerks "in the early days" and serving double duty in the capacity of assistant directors or production assistants who could go "running off to find things."[57] Morris Abrams described the system at MGM in which men filled this hybrid of on-set support roles:

> When I started as a script girl, script bitch, script clerk, "Hey Script," or whatever they called me, I did essentially everything a script supervisor does today except I didn't have to worry about dialog. . . . There was no such thing as a second assistant director at M.G.M. then. The script clerk had to make call sheets, help herd the extras, make production reports, and sometimes, on night calls, pay the extras.[58]

Peggy Robertson herself did not encounter male clerks in her early work in the United Kingdom, but pointed to location shooting as one possible reason men "came into it," in productions when the company was "mostly men, or climbing up mountains or that sort of thing."[59] However, unionization formalized and demarcated divisions between crew positions and defined job duties more strictly. Under this process, the purview of any such hybrid script clerk-camera assistant was limited to clerical work from a chair near the director, because, as Robertson explained, after unionization, "you wouldn't have an assistant director doing script supervisor and vice versa."[60]

By 1921, when Alma Young first stepped into the role of script clerk at Robertson-Cole (which would become FBO, then RKO), continuity workers were typically female, so much so that they were commonly referred to as "script girls." Young's account gives a sense of how, as the title "script girl" implies, the work was implicitly understood as women's work. She recalled,

> I had turned down that sort of job because I didn't pay much attention to what was going on. I had a really lousy memory. But they thought I should keep script. Anyway, the girl that was working for them on that was ill, so I got stuck on it. . . . So three days went by, and the cutter comes by and he wants the cutting notes. I say, "What are cutting notes?" . . . I didn't get fired. The assistant cameraman . . . decided to break me in.[61]

Despite total failure in her first attempt at the role, to the crew, Young was the only real candidate for the job because she was the only "girl" on hand. Her early work at Warner Bros., where she remained from 1923 to 1960, often reflected her gender rather than her specific skills. "They had one [script] girl only there," she recalled, and she was added as a second, but also called upon to step into other "girl" roles: "I would only take an half-hour [for lunch] and then go to relieve the switchboard operator, because there was no-one else to do it. . . . I answered fan mail. I read stories. I was in the property department. I was all over. It was really fantastic."[62] Young played back-up in whatever department needed her help; however, with the possible exception of work in the property department, she seems to have generally been filling in for other women's roles, foremost as the indelibly gendered switchboard operator.

By the time Valeria Beletti set her cap at a script clerk job in 1926 the position was very clearly linked to secretarial roles, though many women saw it as a step up because of its place in physical production. Beletti was unsuccessful in her attempts to get director Henry King to hire her as his combination secretary-script clerk, but many later script supervisors, including Meta Carpenter and Catalina Lawrence, came to the field through secretarial roles. The latter worked at Hal Roach studios as a secretary until she was finally able to land on set.[63] A 1939 profile of the script supervisor's job in the *Warner Club News* describes the job as typically female, but explains that readers shouldn't draw the conclusion that candidates had to be women just because of the article's "constant mention of the feminine sex," adding that at Warner Bros., "there are two very fine script clerks who represent the masculine sex."[64] By the 1950s, Mervyn LeRoy reported that the job was open to men and women but that "two thirds of guild members are women," adding that the job provided "very good opportunity for men to advance to directors, producers, etc. Most women consider this their career, but a few have become writers, actresses."[65] The field has continued to include men, but always in smaller numbers than women, and, of all contemporary media professions, script supervision is perhaps most widely conceived of within production as women's work.

For Peggy Robertson, script supervision seemed destined to be a women's role—even before it included heavy typing requirements—simply because it was thankless detail work. She explained:

> Women are supposed to be, or were, I don't know if they are today, better at detail than men. And whereas the cameraman has a large canvas and the production designer and all these people have the whole set and the glamorous, the script supervisor is glued to the telephone or microphone. The only time that you notice it . . . it's a negative thing . . . when you go and see dailies or rushes, everyone sits down and they look and they say, "what a beautiful job of photography! What a wonderful job of set decoration!" No one ever says

to the continuity girl, "what a wonderful job of matching you did!" They only notice "Why is his cigarette halfway down?" . . . They only notice when you make a mistake.[66]

Unlike other members of production whose work left visible evidence in the finished film, the script supervisor's success was measured by what she kept off of the screen. Through her work's only tangible product—clerical output in the form of notes and typed pages—she ensured against others' mistakes, subtracting the contributions of other personnel that did not match (for example, costume, ad libbed dialogue). Thus, by both design and definition, no script supervisor could proudly point to an onscreen detail as her work, because it only became hers (and not the designer's or the actor's) when it was a mistake. Even in pre-production, Robertson, like others in her profession, was heavily engaged in risk management. If a set were built with a door that opened in a different direction than the exterior it needed to match, for example, although not technically the script supervisor's fault, the error "would be if you hadn't noticed it."[67] A profile of the role aptly titled "Script Supervisor Has No Margin for Error" stated that workers' "specific product" was continuity, which allowed scenes to "seem to flow, as if the shooting had gone on from the beginning of the scene to the end without interruption."[68] Alma Young summed up the job description more succinctly, remarking that script supervisors were "professional fault finders."[69] These and other similar descriptions of the work span the decades, from the 1920s to the present. Indeed, accounts of many other aspects of the work evince striking similarities despite being related almost a century apart, but for certain technological shifts (in both the script supervisor's equipment and the camera and other production technology). This is not to say that there are no differences in how the job was done; simply that in its most essential functions, the work of script supervision has itself shown strong continuity over time.

Though successful script supervisors left no onscreen evidence of their work, a more thorough accounting of the job reveals just how much work it took to make scenes magically "flow" together, and how closely much of it resembled traditional feminized labor. Requirements only increased as production practices and hierarchies developed, and all routine and clerical tasks on set eventually trickled down to the script supervisor. The 1934 edition of *Careers for Women* identified the job as women's work not only because it was one of the motion picture fields most open to women, but also through the description of its duties. The handbook quickly distinguishes the (male) director's creative work from the (female) script clerk's record of it: "Her job is to care for the director's copy of the picture which he is making, to check off each scene as it is photographed and to number and describe the

scenes. She must be extremely careful to make notations of the kinds of garments worn by stars, the physical appearance of the actor, the entrances and
exits and of the details of scenery and finishing."[70] Contemporary and former script supervisors alike use words like *care, worry, anxiety,* and *concern*
in describing their workplace responsibilities. Peggy Robertson, who "kept
script" in the 1940s to 1950s, described her process as a series of steps, starting with reading the script and making notes of continuity concerns within
the text ("How did he get from Iceland to Denmark? If there was something
strange about that, I'd write a memo to the director") or of particular props,
costumes, and action to be concerned about later.[71] She explained, "And you
would be worrying. . . . You'd be making a note of such things as is he carrying a walking cane, is he wearing glasses?"[72]

In production, Robertson's role was to keep track of every important
detail of action, dialogue, and mise-en-scène, such as where a character's
hands were during a particular moment. "That you put in your script, as
much as you can get in, by the dialog, *or* by the action. Anything else, like
wardrobe that has to be matched, and *long* notes, go on the opposite [blank]
page."[73] Additionally, it was her responsibility to make numbered records of
each setup and shot, describing each take (angle, quality of performance, and
so on), when it cut and, when applicable, "why we went again," along with
the time at which it was taken, the number of minutes of screen time (gauged
by a stopwatch) in each take, and any notes from the director. She might type
these notes between takes and, after the day's shooting was ended and the
dailies from the previous day had been watched, was also responsible for
marking up scripts for the editor. A recent edition of a production manual
held the script supervisor responsible not only for wardrobe, dialogue, and
action, but also for recording camera positions and angles as well as "lenses
used, timing, f-stops, and focus changes, such as zooms."[74] Though not specifically cited in studio-era descriptions, existing script supervisors' notes for
specific productions include such notations as lens length, shot distance and
angle, and camera movement.[75] Of course, members of the camera department also recorded camera information, just as wardrobe people kept track
of costuming in each shot, but the script supervisor also tracked it because
they would be blamed for a continuity error even if it was caused by
wardrobe's mistake, as a Warner Bros. script supervisor noted when she
characterized her role as "the original 'fall guy' on the company."[76]

Catalina Lawrence's account of her work from the 1940s to 1970s further
explains the job's tracking and risk management function, saying:

> As soon as I get a script, I read it through to see if I can catch any
> holes in the story. Usually, I catch a lot of them because from the time
> the writer wrote the first version of the script it has had so many drafts

made of it that a story point can get lost. . . . During this phase I also time the script. I take it page by page and act it out using a stopwatch. Then I start my breakdown of the script by making a list of all the important props, when and where they are used, and by what actor. . . . Then I break the script down for wardrobe changes.[77]

Script supervisor Kerry Lyn McKissick, whose credits span the 1980s through the present, picks up where Lawrence left off in her discussion of matching:

I make sure [actors] repeat that action exactly the same way from take to take; if there are differences, I inform the director of them. Ultimately it's the director's decision whether it has to match or not. But it's my job to keep track, so that in editing they're aware of it when they cut from an actor's close-up to their medium shot to their master. Sometimes the rehearsal has taken place months before, so I share my staging and temperament notes with the actors to help them remember the ideas they began with.[78]

Here the script supervisor, not unlike the typist or telephone operator, functions as the conduit through which messages pass from one member of a production team to another or even, in the case of actors returning to roles months after rehearsal, back to themselves.

Like McKissick's, earlier script supervisors' primary concerns involved keeping track, remembering, and making sure on behalf of others. Many found the work stressful, because, as the *Variety* article explained, "Unless everything matches perfectly, the result is not good enough."[79] Peggy Robertson actively "hated" the work and recalled losing sleep to her anxiety.[80] But even those who enjoyed script supervision often described it in terms of worry. And for good reason. "If a line is blown, it's the script supervisor's problem," Morris Abrams explained. Even when other departments might be responsible for the elements that did match, he added, "Basically, the script supervisor is the back-up. The script supervisor must be observant enough to catch the mistake and prevent it from causing trouble."[81] When Lily LaCava supervised continuity on *Gilligan's Island* (1964–1967), an actor performed half of a scene with his hat on his knee and then, after a break, performed the rest with the hat on his head. Despite the responsibility of the actor—and the wardrobe assistant for the incorrect costuming—LaCava said that the next day, when the scene had to be reshot, "There was only one person who got the blame—me."[82] Robertson frequently referred to a similar incident in her oral history, calling it her biggest mistake as a script supervisor. At the end of a day of shooting on

North by Northwest, she noticed actor Leo G. Carroll wearing his own reading glasses in a last take, rather than the tortoise-shell frames he'd been assigned as part of his costume: "I rushed down to see [the dailies] at 7:30 in the morning in the projection room and there's a shot of Leo wearing the wrong glasses. So [the editor] said, 'Well, you've got to tell Hitch.' I said, 'I know, you tell him.' He said, 'No *you're* going to tell him. I'm busy.' "[83] The incident reveals both the stakes of the script supervisor's professional fault finding, and but one characteristic that marks the work as creative service: delivering bad news and the accepting blame for production errors. The group of script supervisors interviewed in "No Margin for Error" agreed that no matter the mistake, "It was really the script supervisor's responsibility." Even if every detail in every department wasn't entirely their responsibility ("No one person can see everything, no matter how observant"), it fell to them to absorb blame.[84] This article's author observed, "With an outsider in their midst, not one would tell anything that might reflect even the faintest discredit on those who had worked on any set with them." One interviewee even insisted her name be redacted after telling a story of a minor wardrobe error that she corrected because "somebody might be able to put the pieces together and figure out from [my name] who the wardrobe man was," leading the reporter to opine that "tact is also part of a script supervisor's professional equipment."[85]

Though akin to the secretary's emotional management and confidence keeping, the script supervisor's psychic support role occasionally required a sharper edge. Much of the work involved telling members of that crew that they were doing something wrong, oftentimes as the only person on set other than the director who was empowered to do so. "Everyone used to make fun of me when I carefully checked all wrists whenever we got ready to shoot," said May Wale Brown, who ensured that no wristwatches made it on camera during the run of the 1800s-set *Bonanza* (1959–1973).[86]

It was not always easy to be the bearer of bad news, pointing out others' mistakes, or being blamed for them herself. Brown didn't relish having to interrupt the flow of a large crowd scene to have the actors put their hats back on to match an exterior shot that followed it chronologically, or catching on to a choreographer's ignorance of the 180-degree rule, only to provoke the ire of director Michael Curtiz, who "jumped down my throat wanting to know why changes were being made when we were ready to shoot."[87]

The script supervisor risked offending her director if she didn't properly package the bad news she was duty bound to deliver. Alma Young did not shy from this responsibility, which may have been why she was sought out for it in an incident not directly in her purview but involving D. Ross Lederman, the director she was working with at the time. He was a "tough, rough guy" known to be difficult sometimes as when he wanted a big explosion in one of the film's scenes. As Young recalled:

> The powder boys arranged the thing so that the next morning they could shoot it. Well, those men know pretty well what they are doing. And late that night they came to my room, and said, "We are very worried. We have just been told that Ross went out there and added dynamite, and if he did, we're in real trouble." So, the next morning, I remember I told the business manager and the assistant director to hold everything, not to shoot until they went with the powder boys to inspect everything. Ross hated my guts from then on, and I had to work on all the pictures that he was directing.[88]

The story reveals the script supervisor's role as perceived by those around her. Concerned crew, intimidated by the responsible party—their boss—approached Young. As the repository for bad production news, responsible for confessing to any slips that might interrupt the creative flow of production, she was the go-to woman in this situation.

Far more often than bringing directors' behavior to the attention of management, Script supervisors were counted on to bring production problems to the director's attention. Perhaps as important as the content of such messages was *how* she delivered them—as his low-status female helper. The designated worriers in production, script supervisors were also responsible for voicing concerns to the most creatively and managerially important person on the set in a way that neither upset him nor challenged his authority, either of which might affect productivity. In an advertisement for AGFA film, entitled "I win a bet from Billie—the script girl," a director's script supervisor performs this balancing act when questioning the lighting of a scene. Billie is quick to manage her status by agreeing to bet him "one steak dinner" on the matter.[89] Though a playful advertisement about fictitious characters, what rings true is the way the script supervisor wisely offsets her difference of opinion with the promise of what—it seems—will be a date with the director regardless of the outcome.

Male script supervisor Bob Gary was no more willing to put himself in overt conflict with a director than were his colleagues of the opposite sex. "Never tell a director he's wrong," he stated, explaining, "Nobody wants to be made to look like a fool, and we all know how easy it would be with a director, who has so much responsibility and might easily be distracted." Instead of telling a director straightforwardly that the camera should have been rooted in one place on a take, Gary might ask, "Did you know . . . that the camera was moving at that time?" He promoted this more passive strategy as a means of bringing attention to a director's mistake while leaving open the possibility that it was really the fault of the cameraman or some other member of production.[90] Peggy Robertson took a similar tack with first-time director Gordon Wellesley:

> *Every* shot that he did was a traveling shot. Panning, zooming, dollying. So the cameraman and I sort of got together and said, "This is going to look horrible, this picture, with the camera zooming right and left." So he said "You tell him." I said, "You tell him." At any rate, I was elected to go and tell Gordon. I said, "There's an awful lot of movements in this picture, we're panning and tracking and dollying. Do you think that's good?"[91]

Delicately phrasing criticism seemed necessary to maintaining respective statuses. May Wale Brown recalled one exceptional occasion when this unwritten rule was broken—on the set of what was then an equally rare film with a female director: *Rabbit Test* (1978), Joan Rivers's directorial debut. On meeting Brown, Rivers exclaimed, "You and I are attached at the hip. Don't ever leave my side during the shooting of this picture!," then insisted that Brown remind her of everything she didn't know, adding, "and I think there is a lot of that."[92]

Script supervisors also served as buffer between a film's director and its cast and crew. Busby Berkeley, who was "rough to work with" and who pushed dancers to their limits, insisted on having Alma Young as his script supervisor for this purpose, because, as Young remembered him saying, "she tells me what a damned ass I am, and so I behave myself."[93] Peggy Robertson tried to maintain good relationships with the assistant director, because "he relies on you. 'What's the next shot going to be? You've talked to the director, where's he going to cut here?"[94] Her employer, Alfred Hitchcock, "hated fighting," and often left her to handle the fallout of on-set controversy that he often caused.[95]

In addition to this job-specific emotion work, many crews expected script supervisors—often their only female members—to perform more general feminine roles. Occasionally, this meant playing hostess, sister, or friend to other women on the set, or to the crew.[96] One Warner script supervisor said that one of her main on-set jobs was "to listen to everyone's tales of woe."[97]

Far more often, though, the female script supervisor served as foil for her all-male crew. Meta Carpenter recalled that crew members often displayed vulgarity that she judged was calculated to shock her as the only woman on the set. Though she reflected fondly on the teasing she received from John Huston and Humphrey Bogart, which she saw as affectionate and inclusive, she more commonly experienced a cruder, more hostile brand of humor on set, which effectively isolated her from the rest of her coworkers. Early on, bolstered by her love affair with William Faulkner, she determined, "Even the narrowness and bigotry of the typical Hollywood union crew— Jew-haters, Roosevelt-cursers, Communist-fearers, denigrators of Catholics and blacks and Mexicans, espousers of the Silver Shirts, opponents of liberal legislation—would not bring down the sky kite that was my heart."[98] However, as time went on, the hostile climate began to have a corrosive effect on her workplace identity. Carpenter described herself, especially when she was younger, as having "a kind of prim, wide-eyed lady-librarian quality that made certain men want to shock me."[99] But rather than playing this feminine role in response to her colleagues, she shifted in a different direction to cope with the performances of masculinity that came at her expense. As she explained, a woman working on a film crew needed a strong personal life to "counteract the mutative process" that occurred when she was mixed in with a group of men who were, with few exceptions,

> power-directed, paranoid, insecure, often sadistic, mulish, and coarse. It is not that these lonely women take on the masculine grain but their womanliness is chipped away in the daily give-and-take with male co-workers whose hostility pours from them like sweat. Her voice unconsciously deepens. Her stride bespeaks efficiency and resolution. Outwardly, she becomes androgynous. The men with whom she works call her by her surname—Jones, Purcell, Lattimer. They tell dirty jokes, ignoring her presence. She has become invisible to them.[100]

Unfortunately, isolation did not mean immunity to the sexual advances of coworkers. Even though she played down her femininity, Carpenter stated, "No passably attractive woman attached to a movie company on a distant location need ever lack for dinner invitations from men or for male bed partners," since she would have offers from actors asking to run lines in their hotel rooms, along with "assistant directors, production managers, cinematographers, camera operators, and sound mixers who slipped off their wedding rings on charter planes after takeoff."[101] As always, it was the woman's responsibility to manage unwanted attention. May Wale Brown violated the unspoken rule of neutralizing male attention while preserving male egos when assigned to director Michael Curtiz. She became concerned when, at their first meeting, he kissed her hand and asked if all her directors fell in love with her. Brown thought, "That's all I need—to have to fight off this lecher for six months." She tried to put an end to it before it began: "With all the sincerity I could muster, I replied, 'Never, Mr. Curtiz.' His eyes filled with anger. Dropping my hand, he curtly dismissed me. . . . I should have kept my mouth shut. During the shooting of the picture Mr. Curtiz never missed a chance to challenge my ability, and he enjoyed trying to ridicule and harass me."[102] The hiring market surely added pressure to "make nice" with superiors on set. Before unionization, many were also expected to work nearly impossible hours (Alma Young reported being twenty-four hours on, eight hours off, seven days a week for months in the 1920s).[103] Eventually, Meta Carpenter joined other script supervisors "sick of second-class status in the motion-picture industry, to organize the Script Supervisors Guild and to win belated recognition from the Producers Association."[104] Even after the formation of their union, International Alliance of Theatrical Stage Employees (IATSE) Local 871 in 1958, script supervisors were not under contract at studios but were hired from picture to picture.[105]

As the field professionalized, script supervisors nonetheless struggled for respect—and to have coworkers and colleagues call them by their job title. In 1947, Jack Hellman reported in his column for *Variety* that a script supervisor had taken the trade paper to task for using the term "script girl." She explained that the script supervisor was more than someone's girl. As the director's right hand, she had "the full responsibility of timing, of making all the script changes in fact, about everything except the directing. . . . I can't see a man in this same position being called a script boy." Hellman seemed to agree, concluding, "Now if some director will only come out big and say the kid's right."[106] Many directors did, yet two decades later, Catalina Lawrence interrupted Mike Steen during their interview for *Hollywood Speaks: An Oral History* to make the same request, saying, "You know, Mike, I use the term 'script girl' and 'script clerk,' but nowadays they prefer to be called script supervisors." Steen agreed that the gender-neutral title was "more fitting, since it is a profession for men also!," missing the point that the word *girl* wasn't just used to distinguish a female from male clerk, but to trivialize the "girl" and her work compared to that of her peers in production, whose titles referred to their jobs, not their gender category.[107]

In game theory, the magic circle defines the space of a game's play. The inside of the circle is distinguished from the outside through such questions as, "What does it mean to enter the system of a game? How is it that play begins and ends? What makes up the boundary of the game?"[108] A similar magic circle might be said to enclose the creative space and its process and players. Clearly, script supervisors stood outside this circle and, through their own labor, created and sustained its boundaries. Though they worked on set—as geographically close as they could be to the process of production—they were segregated by the creative service they performed. Much like female clerical workers elsewhere on the lot, script supervisors supported and facilitated the creativity of their superiors and coworkers through their feminized labor and the womanly role through which they delivered it. They were there to ensure the fun of creative work specifically through their own abstention from it. By minding the filmmaking's details and managing the emotional impact of its difficulties, they assumed the collective worry of production, freeing participants to "play" in the creative space. Script supervisor-turned-production manager Morris Abrams explained how this work benefited the process: "Filmmaking is a business and an art of details. . . . If the little things are handled so that you can work smoothly and have a little sense of momentum on the set, then you get a better film."[109] By overseeing the "art of details," script supervisors served as the set's designated timekeepers, taskmasters, and rule-minders. Most of all, they performed this function for directors, who offloaded responsibility for practical concerns about the continuity of what went into the frame, thus freeing themselves for more artistic ones. This much is clear from frequent references to script supervisors as the director's assistant and second memory. When asked if it was accurate to call the script supervisor a director's "right hand man," Catalina Lawrence echoed the ad's copy, saying, "That's right. Our main job is to keep him straight and not let him forget anything."[110]

Connections to Other Continuity Professions

These characterizations of script supervisors as the director's helper, extra memory, et cetera, hint at the field's relationship to others who performed similar functions in pre- and post-production, and that were female dominated or female friendly in the studio era. Editing, though never fully feminized, has been identified with women throughout film history. Karen Mahar states that the editor role emerged "with the rise of the continuity script and the central-producer system." Prior to that, editing had been the duty of the director. The editor was thus closely affiliated with the continuity writer and script supervisor in that by "using the continuity script and the slate numbers as a guide, the cutter could assemble a rough cut, and even a final cut, often without the director's personal instruction."[111] At the Ince studio, as a film editor detailed in 1922, this practice was assigned to "the girl assemblers" who, before the editor (described with male pronouns) began his work, were given "a layout which they follow[ed], cementing the various strips in their proper order with the titles represented merely by numbers corresponding with numbers thrown on the screen." This intermediate step in the post-production process allowed "the director an opportunity to view his own work . . . making whatever changes he [deemed] necessary and adding any suggestions calculated to improve the titles." Only later did the editor complete the cut, after which it fell again to "many girl assistants" to finish the master.[112]

Editors also performed as the director's "added memory" in continuity and style.[113] According to Edward Dmytryk, in the early years of Paramount, editors were often present on set to get a feel for films prior to the edit.[114] Some directors found cutters a threat to their authority. Others welcomed them because they had no real assistant in production. Dmytryk explained:

> The A.D. (Assistant Director) is an invaluable man, of course, but he is the foreman, the whip hand who, with the help of the production manager, keeps the company operating smoothly. . . . But, in a large percentage of cases, the assistant is really the production department's watchdog, helping but not really working with the director at all. . . . So if a director needs a "bouncing board," he looks to someone else. . . . The cutter, on the other hand, drops into that slot nicely. To the production department, he is always a "director's man." His filmic concerns are close to the director, he has plenty of time on his hands, and he is usually a good listener. He can certainly help solve some simple problems like, "Do we need a close-up here?" with some authority.[115]

Early cutters on the set served much the same function as the script supervisor—right hand, loyal to the director above all, and functioning as his sounding board and auxiliary memory. Viewed in this way, the line of continuity can be seen to extend through the filmmaking process to the script supervisor who maintains that continuity in production and passes it on to the editor in post. In fact, script supervisors had to know a lot about editing to succeed in the role; as Catalina Lawrence said, "We are the link between the director on his set and the film editor in his lab."[116] Not only did the script supervisor ensure that editors would be able to cut films together to make sense through maintaining continuity during shooting, but they also handed off notes to the editor explaining which takes the director liked and any special notes about a given scene. Director, script supervisor, and editor often viewed dailies together, helping both editor and continuity worker further grasp the director's approach.[117]

The connection between these roles—as well as the pre-production role of continuity writer—as partners in "minding details" through production— illuminates their status as female-friendly or female-associated jobs at studios. Many female workers found their way to roles as cutters at early studios from lower-level feminized sectors in the lab in much the same way as secretaries, readers, and other low-rung female workers in scenario departments advanced to roles as continuity writers and screenwriters. Initially, they were able to advance to higher-level editing roles because much of the work surrounding it was considered tedious and routine and thus deemed less desirable by male counterparts. In the 1910s and early 1920s, studios recruited women from the polishing and joining rooms to become "cutter girls," which was how a number of successful female editors entered the field.[118] Margaret Booth recalled, "Irene Morra was the negative cutter and she took me to help her and showed me how to cut."[119] Dmytryk recalled learning about editing's more mechanical, routine labor from the predominantly female staff of the early cutting department at Paramount:

> Hand splicing was a skill, though a minor one. Splicers had to learn just how much of a frame to cut, how to lick the overlapping bit of film with just enough spit to soften the emulsion that had to be removed, how to scrape it off with an Eveready razor blade without weakening the celluloid base underneath, how to apply the right amount of cement and then fit the pieces together so precisely that the doubled film would ride smoothly through the sprockets of the projection machine.[120]

Light manufacturing duties were not the only aspect of editing that set it on an early course toward feminization. In discussions of women's filmmaking specialties from the late 1910s and early 1920s, female editors were cited alongside "woman script clerks" and scenario writers as evidence of women's aptitude for managing detail from positions that didn't require them to exercise physical strength or play a visible leadership role.[121] Not surprisingly, early editing suffered from a lack of respect and acknowledgment in production. Just as some directors felt an editor's presence on set as a challenge to their authority, during the editing process, the editor's role overlapped with the director's. As Robert Karen and editor Ralph Rosenblum explain in *When the Shooting Stops . . . The Cutting Begins*, the editor's level of contribution "came to rest chiefly on how much the director respected and encouraged it," which meant that "outside these two men, almost no one had any idea exactly what that contribution was."[122]

Like script supervision, costuming, and casting—all of which eventually feminized—good editing was largely defined by how little attention it drew to itself. When an editor attempted to assume more creative territory, he risked offending the directors, and had to do so "gently, without causing offense, perhaps even hinting that his innovations had been the director's unspoken wish all along," so that he might be hired for subsequent projects. But, with such limited room to advance, many other editors "learned to play the mechanic's role, sometimes to the extent of a maddening refusal to take any initiative at all."[123] The ambiguity around editing (and what was and wasn't the editor's work) was compounded by the shorthand that attended it. Per Rosenblum and Karen, "'Fill in the holes!' became the great command that editors were left with. An order that had the ring of 'Patch up my grammar when you type the letter.'"[124]

Just as the secretary in the office functioned as the executive's or producer's proxy, often making important creative and managerial decisions, the early editor's success came through willingness to minimize his or her own achievements. According to these unwritten rules, an editor could accept credit for the more rational, mechanical aspects of assembling the film without taking credit for any of the artistry that makes a story flow seamlessly or underscores its thematic or stylistic elements. And, because of the ambiguity around their work, editors, like script supervisors, could easily be blamed for others' mistakes, simply through their inability to catch, disguise, or erase them. For this reason, it comes as little surprise that women were channeled toward editing from existing feminized labor sectors in the 1920s when "several women graduated into film editing from jobs as negative cutters, script girls, or secretaries."[125] As Rosenblum and Karen explain, women knew how to play the helper role: "Trained from childhood to think of themselves as assistants rather than originators, they found in editing a safe outlet for their genius—and directors found in them the ideal combination of aptitude and submission."[126] Barbara McLean cited women's traditional roles as nurturers for helping them to cut films "like a mother would, with affection and understanding and tolerance."[127]

Women's other "natural" feminine qualities also came in handy because, as editor and director Elmo Williams stated, "They are sensitive and as a rule have more patience than men do"—and, of course, "Women don't mind all the fiddly little details that you have to deal with; they're very thorough."[128] These descriptions of editing highlight its similarities not only with script supervision, but also with continuity writing. Jeanie MacPherson's treatise on continuity writing for a 1922 issue of *Photoplay* is full of references to screenwriters as artists and to continuity writers as architects, builders, and technicians who know the rules around continuity writing. Looking at a continuity script, MacPherson wrote, "The amateur might think that 'it certainly LOOKS easy enough.' So he writes his 'continuity' and behold! It LOOKS just like the work of an expert! It contains the required ingredients." But this amateur would soon discover that his work would not "pass the strict 'building laws' of dramatic construction."[129]

Although each director was responsible for creating his artistic vision, unified by the common job creed of safeguarding a picture's structural soundness, these three roles maintained the coherence of that vision. From the continuity writer who ensured that the story made sense on the page to the script supervisor who minded its pesky details in production, and, finally, the editor who arranged it into a cohesive whole, this tripartite continuity process wove rationality into the fabric of production. Much as MacPherson observed of continuity writing, these cohesion experts collectively served as architect to the director's artist. Each of these jobs constituted a form of creative service to production in general, and to the director specifically, with success often determined by the extent to which they could perform as the director's right hand without causing him to lose any authority—real or imagined—as the film's creative center. Rather than the director's second memory, these professions might more accurately be characterized as his second brain, responsible as they were for much of the complex, practical thinking needed for an artistic vision to cohere as a comprehensible story. Where movies were concerned, God was in the details. And so were women.

The continuity worker was also a resource for the director, offering a second creative perspective under the nonthreatening guise of assistance or service to his vision. This function was especially true for editors, who were responsible for a great degree of the artistry with little credit. Editing resembles the creative service of the studio secretary in this respect, with the addition of significant organizational responsibilities that support the director's process. Editors were responsible for assembling films to allow directors to make key decisions. Like the secretary, the editor arranged options, assisting a creative decision maker by narrowing a field of editing possibilities to a choice few. Editors, obligated to fly their agency under the director's radar, also delimited the solution to creative problems by offering options and coaxing the movie maker in the right direction rather than telling him directly what he should do.

Meeting editing's inventory of requirements, women in the developing industry made inroads in the profession. In the early days, Adrienne Fazan remembered, "Every studio had a few women editors. . . . A woman could get started then."[130] According to Karen Mahar, "At least a dozen other women were counted among the first editors in Hollywood, among them Anne Bauchens, Blanche Sewell, Anne McKnight, Barbara McLean, Alma MacCrory, Nan Heron, and Anna Spiegel."[131] However, despite the early association of women with editing, the field's creative importance did not stay hidden for long. Unlike most feminized fields, the work of the editor was done on film, not paper. And, as such, unlike secretaries, script supervisors, or continuity writers, editors interacted more directly with the creative product. By the early 1920s, editing's visibility had been raised by artistic successes like the films of D. W. Griffith, which owed much of their impact to parallel montage and other new editorial storytelling techniques.[132]

And, as cutting rose in desirability and prestige, male workers staked their claim in editing departments. As usual, the processes of standardization and specialization aided male aspirants. Female editors encountered more male underlings as the loose systems under which they had been hired gave way to a more rigid hierarchy of apprentices and masters into which studios funneled male workers. Dmytryk said that with the advent of sound, feminized cutting rooms underwent "a qualitative and quantitative change." Female cutters were often replaced by department heads who thought they couldn't cope with the new technology. Editing departments also expanded because of the intricacies of sound, which, as Dmytryk explained, "necessitated the establishment of a new working classification—the assistant cutter. There had always been apprentices, but the assistant cutter was a rarity before sound. Now there was a stream of personnel flowing into the cutting department, and because of the real and imagined difficulties involved in cutting sound, that stream was almost exclusively male. Fortunately, I was a man."[133] Dmytryk received his own promotion to cutting when a slot opened up. Some female editors like Violet Lawrence and Anne Bauchens, who had entered the profession in earlier years, remained and had long careers.[134] Margaret Booth served as MGM's supervising editor and reviewed dailies and cuts for every film there, giving notes to the editors under her.[135]

But willingness to allow new women into the field had ebbed. Unions made it increasingly difficult to break into the business as studios promoted from within, which meant they mostly promoted men.[136] As Mahar explains, women already in the field were also affected by the shift toward masculinization: "Even female editors who began their careers in the 1910s and early 1920s ran into hostility from male editors. Viola Lawrence's husband, Frank, who taught her to cut film in 1915 was 'mean' to the female assistant editors he supervised at Paramount in the 1920s. 'He just hated them,' she claimed. 'If any of the girls were cutting—if they did get the chance to cut—he'd put them right back as assistants,' but he 'broke in a lot of the boys.'"[137] When Dede Allen "wormed her way into the cutting room" at Columbia Pictures in the 1940s, she saw the job as a more realistic goal than her true aspiration to direct—where the obstacles were even greater. But that did not mean she experienced editing as a sector that welcomed women. Allen never left work when her children were sick, because she knew her male peers would judge her for it. She tried not to feel irritated "at being a woman, because sometimes you have a lot of gaffe [sic] that goes with it," and attributed her success to being told "You can't" so frequently that "you had to make it happen."[138]

Though editing was not feminized during the studio era, its trajectory toward and away from feminization demonstrates how notions of women's work shaped studio practices and production cultures, and vice versa. When jobs in the film industry were deemed undesirable by men for the same reasons as they had been in other professions—perceived service characteristics, clerical/organizational or rationalized duties—women were allowed in, often on the condition that they took less pay and offered more of the service work that came "naturally" to them.

Women's success in these roles had nothing to do with their gender and everything to do with their individual talents and collective determination to succeed in any job that brought them closer to the same creative, managerial, or financial gains any film worker desired. They would have been equally dedicated in directing, producing, and cinematography if they had had genuine opportunities there, which, beyond the motivation to hire them, would have required that others were willing to train them and accept their presence in the field. Because they were deemed natural editors early on, women's work in editing was legitimized and their successes tolerated; in other, more contested, areas, they were either denied the chance to work altogether, or had their efforts dismissed, sabotaged, or judged on a stricter scale than men.

Tolerance did not equal support or respect. Even the feminized job of script supervisor was ostracized in production, the boundaries between its supporting role and the more creative or technical areas of production policed through work culture. Joan Rivers's request that female coworker May Wale Brown never leave her side as she directed her first film was probably not gender blind. Women could not access the pipeline that trained them for technical crafts like cinematography or directing. If, by some other means, they obtained a directing position, they could expect to fend for themselves and be blamed for the knowledge they didn't have and weren't offered by peers.

And so they frequently took the only paths available to them, those like editing, script supervision, and continuity writing, where their presence was sanctioned by gendered logic, not equanimity or recognition of their individual talents. This meant that when a job's creative capital was raised, women in the field could expect to be marginalized in much the same way. In the case of editing, outcomes were mixed. Some women stayed in the field after the 1920s, but others were pushed out, and many would-be female editors never even broke in. Today, the field retains some of its association with women despite its much-elevated creative status, yet women have never regained their early toehold there, and the profession can be considered mixed gender at best. How different might the field be today if it had continued its path toward feminization? Would women dominate? If so, would the creative status or compensation be reduced? As the final pair of women's film professions demonstrate, the past is never truly the past. Even in a field that had previously been masculinized, was of relatively high status, and had followed a course to feminization that was the inverse of editing, when women stepped in, the ground shifted under their feet.

Publicists and Casting Directors: The Rule and the Exception in Late Feminization

The feminization of casting and publicity departments began late in the studio era—as male managers promoted women out of purely clerical positions to planning positions above them—and was completed after the big studio period was over.[139] Unlike the script supervisor, who could not cross the culturally enforced gender barrier between her "soft," paper-based skills and the "hard" technical expertise of nearby masculinized positions, women in planning departments benefitted from their proximity to processes that took place on paper. But these benefits were slow to manifest.

Publicity and casting had been female-friendly occupations in the early industry and were identified as such by Myrtle Gebhart when she listed them among the professions open to women in 1923.[140] Around the same time, Lasky publicity director Adam Hull Shirk wrote:

> Women as well as men work in the publicity departments of some companies, though the latter are in the majority. Sometimes, especially in the matter of fashion articles and the more intimate details of feminine life, a woman is better qualified than a man, but many feminine readers of stories concerning the sartorial characteristics of the stars, would be surprised to learn how often those articles are written by mere men who have caught the knack of description and learned how to intrigue the interest of women readers.[141]

As the industry grew, masculinization claimed both publicity and casting, where women were largely excluded from leadership positions—and much of the work underneath them. As with other professions on the lot, it was easier for men to move into publicity jobs because they had more direct access to other roles at studios. Shirk advised that if he had no other "in" to publicity, he would simply get a job on the lot and hope to learn the business while positioning himself to ask for an opportunity when it arose.[142] There were simply more jobs from which men might accomplish such maneuvering. Ronald Davis writes that most unit publicists were male prior to World War II, which was "considered appropriate at the time since the job required traveling with the company while the production was being shot," and publicity departments in general were male dominated and headed by men through the 1930s.[143]

Women did have one important "in" as secretarial workers in the department, which allowed them to eventually regain ground there. MGM publicist Howard Strickling was "one of the first to hire women," and the field integrated relatively steadily once the practice began.[144] The leap from clerical/administrative duties in publicity to actual publicity work was fairly short. Robert Vogel, who oversaw foreign publicity at MGM from the 1930s through the 1970s, described the work of his assistant Peggy O'Day as very close to his own, saying she was really more of a "Deputy Mayor" than a standard assistant, in that she did "everything. Anything that I didn't handle personally she handled. When I was away, she ran the place."[145]

It also required less imagination to accept women in publicity roles, not only because of the paper-based nature of the work, but also because of the job's requirements and perceptions thereof. Women's presumed skill set could more readily be framed as an asset to publicity because the work involved behind-the-scenes writing—reporting about films and stars in departments that functioned a bit like newspapers—as well as elements of storytelling, service characteristics such as caretaking (of studios' stars), and sales (of stories to executives and the press), and, most womanly of all: gossip. For all of these reasons, publicist roles were not the ultimate goal for many of the department's male workers. And Christopher Finch and Linda Rosenkrantz note that "it was not unusual for publicity men to move into other branches of the movie industry."[146] This evacuation could only have helped female aspirants, freeing up as it did more spots open for them. Eventually, female publicists integrated even the upper-level positions at studios, and by the 1950s and 1960s were branching out on their own; the job was well on the way to its contemporary feminized and female-dominated state.

I raise the example of publicists briefly here because of its similarity to casting in terms of its planning function, place in studio organization, and the high volume of paperwork it produced. Given these characteristics, under the logic of women's work at studios, publicity's path to feminiza-tion seems almost a matter of course. And, it might reasonably be assumed that other paper planning professions would have followed a similar path. However, casting—which took place on paper and was located near pub-licity in studio geography and hierarchy—did not truly feminize until the post-studio era. The role of casting director was typically filled by men until the 1950s and continued to be male dominated until the 1970s. And unlike other, related feminized sectors, casting carries relatively high creative and managerial status in contemporary below-the-line production hierarchies. Moreover, the kinds of creativity and leadership that are cited as modern casting's "best practices" defy simple categorization as originating in either historically feminized labor sectors or early twentieth-century manage-ment traditions that studio-era casting directors and executives embodied. Indeed, rather than fitting neatly into any one historical explanation, cast-ing is a mixture of the two, with some feminized aspects, some managerial aspects that descend from the scientific management tradition, and many others that are unlikely hybrids of the two. For this reason, constructing a history of the field's feminization requires some unpacking in the present before its traits can be located in the past. The profession's indirect path to female domination demonstrates the persistence of creative service as a condition for women's success, as well as the multiple, often fractured identities today's casting director's embody through their roles in creative industries' complex work systems. As such, casting represents the outer ring of the ripple effect begun years earlier when notions of certain types of film work as women's work first emerged.

Casting's Conundrum: Modern Casting

In contemporary Hollywood, films and television programs are created through an interlocking series of soft systems held together by multiple, contradictory industrial mythologies, resulting in production processes that are often messy, disconnected, and chaotic.[147] Modern-day casting, and in particular its feminized status, exemplifies the messiness. Press and trade profiles of the profession tend to focus on the high ratio of female to male casting directors.[148] This angle provides a sense of historical symmetry, evincing shifts from the old Hollywood values symbolized by "the casting couch," a euphemism for the exploitative practices for which casting was known in the past. However, the profession's gendered state goes beyond having a simple female majority and is more indicative of regressive than of progressive values. Today, for both female and male casting directors, such gender binaries play a persistent role in their understanding of the field, especially because many believe—per professional lore—that the job has been gendered in this way since the early days of film production.

Contemporary casting directors frequently attribute their field's female-dominated state to its considerable clerical-organizational components. Daily work in casting offices includes heavy helpings of such clerical-secretarial basics as "opening envelopes and answering telephone calls and Xeroxing sides and calling and setting up appointments," as well as organizational tasks involved in narrowing the field of actors for various roles, such as filing materials on different performers, checking their availability to work, and typing lists of top candidates.[149] In the oral history passed from casting director to casting director, the profession has "always been a women's field," precisely because so much of the work is clerical "drone work" of the type traditionally associated with women.[150] These accounts maintain that either "it used to be the secretaries who took care of [casting] in the old studio system," or women were assistants to early casting directors, and "were cheaper replacements when the men went on to more lucrative things like producing," because "we can type up our own lists and make a deal at the same time."[151] Though this history isn't completely accurate, its link between casting's female dominion and clerical aspects is.

In addition to the field's clerical elements, contemporary casting directors attribute women's dominance to emotional aspects of the work.[152] Instinct, emotional intelligence and intuition (rather than reason, intellect and logic) are often cited as key to female casting directors' success in the field, because "women frequently have good instincts for casting," for which actors had chemistry, and for which story was the most interesting to tell.[153] Many casting directors assert this talent comes easier to women because "for some reason women are homed in on those instincts . . . about relationships, about people."[154]

Communication skills are also vital to success in casting, with its multiple sessions involving hundreds of potential hires and creative players, requiring much back-and-forth relay of information. Though communication can certainly be classified as a gender-neutral skill, in casting, communication comprises the same emotional labor and service characteristics (pleasant phone manners, people skills, detail oriented, multitasking, forging personal connections) women were expected to bring to the professional sphere from their earliest forays there. In the casting directors' oral history, it follows that women made inroads in the field because "we tend to be a little more natural communicators," which "evolved into having a place for women," who are able to do more than one thing at a time, in a job that requires care for details and work in groups because "women like company."[155] The received wisdom is that women are better suited to the social aspects of what is "a more people-oriented profession," because, "you have to be enormously interested in people, to the extent that you put your own ego aside. Women are trained to do that, to listen and to be very interested in all kinds of people."[156]

Emotion work and service characteristics pervade descriptions of casting sessions, when expectations and emotions of dozens of actors come into play, as do those of a project's directors, producers, and executives. In these sessions, casting directors carefully manage tension and set the emotional tone. Casting directors agree that it is crucial in these sessions "that actors feel very comfortable, that directors feel very comfortable, that you tear down as many barriers between them as possible," while making sure "that everybody feels they are part of it."[157] Indeed, "keeping everybody happy at the same time" is regarded as one of the most difficult aspects of the work because, in addition to pleasing the director, "you might have one, five, seven, twelve people who will all participate in the choice about who this actor is going to be."[158]

Interviewees make frequent comparisons between their roles in casting meetings and women's traditional roles in the home. The work is likened to home entertaining ("It's almost like women have a genetic hostess gene"), a skill that aided them in sessions, where "sometimes you do feel like a hostess at a great party," "introducing the directors . . . making the actor comfortable in your home," "keeping the room alive, getting everyone excited about the next actor, lifting the spirits of the director if 20 people pass on the part. You're like a good wife in that respect. You make sure everyone gets what they want for dinner."[159] Casting directors cite other ways in which the job is like being a wife or a mother, especially in terms of the caretaking involved, saying, "The better casting directors will nurture actors so that they're comfortable in the room," while "casting directors who do not make people feel that they're well taken care of don't tend to work with those people ever again."[160]

As important as it is in casting sessions ("We're very emotional in the casting. We cry in the room"), emotional labor is equally important afterward, when delivering good and bad news is, again, a matter of emotional management, in that "you have to tell people they did or didn't get parts," and though "it can get you down having to say 'no' to so many people so often," good casting directors are those who limit the emotional impact on others by understanding that "there's a politic way to deliver news without offending sensitive people. You have to learn who has a thin skin."[161] Taken alone, these characterizations paint the job as largely dependent on reflecting traditional gender norms. Vicki Mayer has observed a tendency among reality television "casters"—casting directors' nonfiction counterparts— to "internalize a binary logic around gender and sexuality, emphasizing organic or natural bases for their talent" (rather than training or job skills more associated with commerce), which has the effect of devaluing their labor, "undermining its skill set in comparison with jobs that required certification or the registration of formal education on a resume."[162]

Though casting directors similarly cite these "natural" talents, unlike Mayer's subjects, they also emphasize the importance of casting skills associated with the masculinized business tradition, such as deal making, negotiation, budgeting, and pitching their services to new clients. This emphasis perhaps reflects the longer history of the film and television casting director, and its practitioners' recent efforts to correct the perceived lack of understanding of casting and the professionalism of casting directors outside the field. So while discourse around their work often begins with the headline of female authority and women's skills, many casting directors are quick to point out the buried lede: that their work requires many more "hard" skills in the economic and political juggling required to "make the budget work by delivering actors who satisfy the requirements of the roles while bringing a certain panache to the project."[163] Indeed, casting workers frequently highlight the complexity of the deal-making process, which starts with budgeting.

As such, casting directors must have a nuanced understanding of production that extends well beyond their part in it, and know "a tremendous amount about deals. . . . How do I work it out with an actor who I want but has got another movie to figure out how they can do both jobs? How do I . . . manipulate my budget so I can get this actor?"[164] They must also be on familiar terms with the legalities of putting together a deal memo, which is "political on both sides."[165] Other hard requirements cited for casting success were aptitude for pitching casting services to prospective clients, and of course, managing casting associates, assistants, and interns, and to newly hired actors before their first work calls, relaying scheduling and wardrobe instructions.[166]

These managerial duties (financial and organizational planning) align with more traditional notions of management than with the emotional management required in casting sessions, representing a characteristic in common with the work of studio-era casting directors. However, the similarities end there. For, even though the basic, one-line job description (matching actors with roles) is the same today, earlier casting directors were more closely aligned with executives than production crew, with whom contemporary casting directors closely identify.

Early Casting: Asset Management (1900s–1910s)

Early casting practice was typically a modified version of theatrical casting, in which actors were known and hired for "lines of business" or character types they'd mastered (for example, the "heavy") and became through costumes, makeup, and performance. But film's realism and the advent of the close-up made it obvious that actors in costume were just that, and so, fairly early on, film casting was adapted to identify stars and categorize actors by physical rather than character type. Said one later studio manager, the casting director "must be able to pick types who look and act the part naturally. If the story calls for a weakling, he must pick a man with a weak face."[167] "Typage," as the practice was known, dovetailed nicely with developing efficiency practices, by which casting quickly shifted from something done by a cameraman or director—selecting "the leads from his stock company and the extras from anyone who appeared at studio 'bull pens'"—to a standardized system.[168] Extra Charles Graham described this process in 1912:

> We joined a crowd of people. . . . We had not said a word to a soul and no one had questioned us, when a man in shirt sleeves and with a green shade over his eyes came into the room and scrutinized first one and then the other. . . . "I can use you," said he, and handed each of us a card. . . . We learned that the film would be known by this

number til its name was revealed to a waiting public, that we were the "walking gents" in a card playing scene which was to be shot that morning and that we were to take the card to the wardrobe room.[169]

The Premier Film Company hired theatrical agent John W. Mitchell, whose job it would be "to meet and interview all applicants for parts and to tabulate the result of such interviews according to type, dramatic ability and physical qualifications."[170] Other firms hired talent scouts from vaudeville and created positions of casting directors with such directives as, in the case of Equitable's general casting director, engaging "no less than two thousand people a week for the seven companies now actively engaged at the various Equitable studios."[171] Under such "casting efficiency," actors simply became another asset to be tracked as casting directors, in Mary Pickford's assessment of their role, "divided humanity in sections."[172] Classifications were assigned, recorded, and cross-indexed by early casting workers, essentially locking each actor into a specific type that was noted in their records for ease in distributing them to various productions.[173] By 1922 Goldwyn casting director Robert B. McIntyre claimed that his office files contained "thousands of faces, and records even more detailed than the criminologists'—so that when a visualized face appears on the mindscreen of the casting director, he can at once secure this data by naming the person or the type, and consulting his files."[174] The same year, Melvin Riddle described the process and its extensive systematization as it had developed at Paramount: "For every principal, free-lance, and extra player there is a big card with figures giving his or her height, weight, and other physical data. These are cross-indexed into files of types, segregating heavies, juveniles, character people, leading women, leading men, etc."[175] When project planning began, Riddle continued, the casting director went through his files and chose "a leading man and if the latter is available, [put] him down for the part, and so on with the other players."[176]

Through typage and efficiency, studios systematized the process by which they acquired and retained actors as studio properties just as they did scripts and equipment, developing the contract system to lock them into place and assure their availability. H. O. Davis, in describing Universal's system, registered actors more as assets than people, saying, "The director then, in conference with the head of the casting department and the manager of production, casts his picture from our stock (we carry about 300 actors and actresses of various types on the payroll and in stock at all times)."[177] Lasky casting director L. M. Goodwin described his "stock" in similar terms, claiming, "We maintain at all times, a sort of reserve of about a hundred extra people—boys, girls, men and women, who have tried and proved, who we know can work, who have satisfied the directors and who can be depended upon."[178] The Motion Picture Producers and Distributors Association's establishment of the Central Casting Corporation in 1925 as a hiring agency for all productions' background actors further systematized the process. As Anthony Slide explains, Central Casting "was able to provide any type of extra—man, woman, child, or handicapped—and through the forties, extras were divided by Central Casting into four distinct types: 'atmosphere,' 'character,' 'specialized,' and 'dress.'"[179]

Studio-Era Casting: Casting Executives and Talent Groomers (1930s–1940s)

By the 1930s, studios had taken their control of actor assets to new heights, devising "option contracts" that tied actors to one studio term of seven years, giving studios exclusive rights over the actors' images and services, so that they could be required to perform in whatever roles the studio chose and to participate in publicity and advertising. The contracts gave studios the option of renewing or dropping the actor every six months depending on their progress, but didn't provide the same option to the actor.[180] In this rigid system of star control, the casting director was the primary manager of studios' actor assets, stockpiling stars and distributing them to production. Typecasting reduced actors to their age, race, gender, and physical characteristics, and much of the casting process was completed without the involvement of the actors who ultimately wound up in roles. As Beth Day explained, "Once a story had been agreed upon, the producer, director, an executive from the front office, and the studio casting director met to discuss the cast."[181] Though most studios had a dedicated front office executive who oversaw talent and casting, CEOs, studio bosses, heads of production, and other high-ranking executives might also participate in this process. At MGM in the 1930s and 1940s, casting director Bill Grady "cast many MGM pictures in conjunction with [vice president] Benny Thau, who in turn consulted Louis B. Mayer."[182] Lew Schreiber served the same function as Grady for 20th Century Fox and Rufus Le Maire held an equivalent role at Universal, while at Warner Bros. it was Max Arnow and later Phil Friedman in conjunction with executive Steve Trilling.[183]

Ideally, principal and supporting roles would be filled with the studio's contract players since, as Ruth Burch explained, "Every studio had a substantial list of good talent."[184] The Warner casting department notified producers about screen tests of newly signed actors, and once a script had been written, its producer and casting director held a casting conference "to see who [was] available, in or out of the studio, and at what price."[185] Often well-known actors were cast without reading or testing at all. Newer players "were tested for parts through interviews with the casting director, then by a screen test."[186]

Although contemporary casting directors' describe duties that are almost exclusively informational and interpersonal, the accounts given suggest that studio-era casting directors functioned as decisional managers, and more specifically, a subset thereof known by organizational theorists as resource allocators: managers who decide how to distribute an organization's people, time, equipment, and so on.[187] As mid-upper-level managers, studio casting directors acquired and allocated actors strategically in conjunction with top managers from the front office, not on a picture-by-picture basis, but across entire studio slates.

The casting director's manager-executive status was reflected in casting's typical spatial separation from production, both geographically, and in terms of studio workflow and hierarchy, where it was typically grouped with the studio's legal and executive branches—the center of studio planning and management. In keeping with sex segregation practices, a mostly female workforce typed and maintained the kind of clerical output—casting lists, meeting notes, memos—that was casting's physical product.[188] Throughout the 1930s and 1940s, men occupied the managerial role of casting director almost exclusively, and dominated the related positions of casting associate and assistant. Casting offices were often housed near offices of other planning departments such as publicity and advertising, all of which were also headed up by male executives and supported by a predominantly female clerical staff.[189] For example, during his years as head of casting, MGM's Bill Grady "had a five-room bungalow as his headquarters. His own office was protected by two secretaries, double sets of doors, a private switchboard, and a window of one-way glass."[190]

James S. Ettema uses the concept of "players-in-position" to describe media workers whose roles situate them in a position "to participate in the decision-making of the organization," based on hierarchy, expertise, control of information, and access to channels of agency such as negotiation.[191] By this definition, studio-era casting's players-in-position were casting executives and casting directors (who often functioned as executives) along with moguls, other executives, producers, and important directors. However, although they negotiate on behalf of the players-in-position in their productions, contemporary casting directors do not themselves identify as powerful enough to be considered players themselves. Instead, they frequently mention their position's lack of direct decisional power with remarks such as, "Actors think we have that power. . . . But we aren't the ones who make the decisions on who gets hired,"[192] and "It's all really the director's choice. . . . In the end, we're invisible."[193]

Contemporary casting directors limn the subterranean nature of their authority through reference to their oblique power to coax decision makers toward a decision without telling them what to do. One casting director described the multiple vectors of authority in the process: "There is a LOT of psychology involved in handling the large groups of the creative team (producers/writer/director/executives at the studio). You want them to hire 'your guy' and you have to get them to feel that it was their idea in the first place!"[194] Others described delimiting the solution to the casting equation rather than devising it themselves: "I'll try to read 30 or 40 people for any decent sized role and whittle it down to 5 to 10 for the director," and "The only power we have . . . is to tell an actor, 'No, you can't go in to see the director.'"[195] One summed up the delicate process of subverting one's own opinion, managing one's own status, and "imprinting your own taste on the project by who you bring in." As she explained, "You see what you think is right in fact, but you can't go in and tell people. . . . I have to make [them] feel good about a decision. . . . I think it's easier for women to kind of throw back their own vision and sort of nurture people into [a decision, rather than . . .] be heavy-handed about my ego needing to say 'This is it.'"[196] The exception to this push–pull dynamic can be seen with casting executives at studios and networks. Such positions as executive vice president of casting survived at studios throughout the 1950s and 1960s, when the casting director position began to diverge from its previous, executive-managerial identity. Casting executives, who today oversee all casting for their studio or network and manage the various freelance casting directors hired to cast individual productions, are much more closely aligned with studio-era casting directors, overseeing decision making at the level of a player-in-position. These executives do not typically work as closely on individual projects as contemporary casting directors do. Instead, they function similarly to the other executives in the casting process.

Post-Studio-Era Casting (1950s–1980s)

As the studio era drew to a close, increasing numbers of female clerical workers received limited promotions in casting, as well as other planning departments whose main products were paper based, especially in publicity.[197] However, with rare notable exceptions, these workers did not rise to the level of manager or head of their departments. Other than Ruth Burch, who described being promoted to casting director by Hal Roach in the 1930s or 1940s, few women ascended to that level before the 1950s, when Marion Dougherty and others report promotion to casting director, often in new TV divisions.[198] This lineage does little to explain how, between the 1950s and the late 1970s, casting became not only gender integrated but also heavily female dominated, or how feminized duties became not just an added value provided by female casting directors, but their primary value and a means through which they operated creatively.

In reality, contemporary casting's feminized service and emotional labor aspects existed in the studio era, just not in casting departments. Before casting directors could place actors in the roles in which the public would come to know them, an army of other workers was necessary, not just to discover actor assets and lock them into contracts, but also to develop these assets, once secured, so they would be ready to distribute to productions. Whereas the scouts who brought undiscovered actors from all over the country to the attention of studios tended to be male, many of the other jobs under the banner of talent went to women. Lillian Burns supervised talent training as drama coach at MGM in the 1930s and "could end careers, but nurtured those she thought had star potential." Sophie Rosenstein was "talent coordinator, teacher, and mother confessor to young contract players first at Warner Bros., then at Universal."[199] At Paramount, Phyllis Loughton served as drama coach, while Helena Sorrell led talent at 20th Century Fox, and RKO employed Ginger Rogers's mother, Lela Rogers, for the same purpose. After the executive decision was made to acquire an actor asset, these drama coaches and other female studio caretakers nurtured and developed them, both personally and professionally.[200] Young contract players underwent an extensive apprenticeship program to prepare for stardom. Ronald Davis described Lillian Burns's process:

> Most of her studio contact was with Benny Thau and the producers and directors of specific pictures, although she regularly conferred with talent scouts and the casting director, Billy Grady. . . . She worked with her contract players for weeks on specific roles, preparing them more thoroughly if she knew they would face a weak director on the set. She would take a young person through an entire script, working as she would in a rehearsal. But a major part of her task was developing the players as people.[201]

MGM and other studios invested in elaborate processes of training and grooming to render actors a total marketable package, from acting, dance, movement, diction, riding, fencing lessons, down to the selection of their clothes and, if they were successful enough, setting up households in keeping with their image.[202] Drama coaches read studio projects with new hires in mind and brought them to the attention of producers and executives who could place them once they were ready.[203] Unlike casting directors, talent workers made daily use of the female-associated skills that today's casting directors describe, such as intuiting an actor's "rightness" or "readiness," nurturing actors in and out of auditions, serving as acting teachers, participating in the decision-making process through influence and solution delimitation—rather than direct commands—and mitigating the emotional content of messages during the casting and talent grooming process. Here, again, the studio made use of women's labor in specific, targeted ways. Here, again, women added value to their labor through feminized duties. Though talent coaches and casting directors did very different jobs in the studio era, when the contract era ended and both actors and casting directors began to work as freelancers, many of the competencies that drama coaches and talent groomers used were shifted over to the casting director's job description.

In the economic downturn that followed the 1948 Paramount consent decree and forced divestiture, downsizing studios turned to newly independent producers to assemble film packages ad hoc, gradually ending the contract system for most talent, crew and craft workers, outsourcing much of the production labor, and hiring most movie workers from pre- to post-production on a per-project basis.[204] The new freelance system removed the guarantee of steady employment that had existed for formerly contracted workers at studios. Smaller firms sprang up around various aspects of production (from craft service to sound mixing), ensuring their survival through "flexible specialization," a strategy that customized work processes to fit the individual needs of various clients on a project-by-project basis.[205] Though there were casting directors on staff at studios throughout the 1950s and 1960s, the need for them declined as the number of players under contract dwindled. By the mid-1970s, casting had become largely outsourced to independent casting directors such as Lynn Stahlmaster and Mike Fenton, who had begun to form their own firms for the purpose.[206]

In its freelance incarnation, casting struggled to reconcile studio-era practices with a new economic model. Where previously it had been standard practice, "for the producer and director to sit down with the casting director and cast a film in 20 minutes," studios' economies of scale were changing.[207] No longer able to rely on talent scouts to find, test, develop, and secure a stable of supporting players and stars, freelance casting directors began to work a pool of actors reaching into the tens of thousands, including freelance stars who now had to be courted and hired rather than simply assigned as they were in studio days.[208] Ruth Burch described the increasingly complex process in 1969, saying, "There are now over one hundred and fifty agents in this area. As you deal with them, you learn which type of talent they handle and which of them will cooperate with you in negotiating a price or which will hold fast."[209]

This more anonymous process initially veered toward tried-and-true efficiency techniques to narrow the field and cast dependable, agreed-upon types in roles quickly, as was often required by television production schedules. Describing TV casting in the 1970s, Joseph Turow wrote, "Although casters who were interviewed use different personal filing systems, all those systems are aimed at associating actors with parts for which they were cho-sen and suggested in the past. This procedure ensures the caster's ability to choose quickly and credibly; it also results in patterned looks for certain parts."[210] Early female freelancers, like their male counterparts, modeled their process after that of studio-era casting directors and were, according to one casting director working at that time, "much more about dealmaking, much less about artistry than they are today."[211]

However, the economic shift coincided with—and was influenced by—a creative one. Studio-era producers and executives controlled the casting process, and although "prominent directors might also play a role in cast-ing," outside of those roles in which unsigned newcomers might be cast, they were generally restricted to actors under contract or who could easily be lent from another studio.[212] But New Hollywood sought less generic, more organic casting. A new generation of film directors came to the fore, empha-sizing location shooting and grittier, more "authentic" stories, aesthetics, and actors. Freelance casting directors—and a vanguard of female trailblazers in particular—distinguished themselves from their studio-era counterparts along similar artistic and creative lines.

Marion Dougherty characterized the Old Hollywood casting director as "a grocery-list maker [who] believed casting meant that if you make a list of everybody from Shirley Temple to Tallulah Bankhead, somewhere in between you had to have somebody who was right for the part."[213] Joyce Selznick, whom Harry Cohn appointed to run his East Coast talent and story departments in 1941 and who served as worldwide head of talent for Paramount in the 1960s, also saw a creative separation between old and new casting methods:

> There are casting directors who submit lists of people whom they are familiar with and leave casting decisions pretty much to the producer or director. They take a rather passive course. On the other hand . . . I cast a picture a few years ago, and in reading the screenplay I saw that the leading character wasn't finely etched. I thought we should not cast it in the way it was written, because if we did we were going to end up with exactly what we had on paper: an uninteresting character.[214]

Increasingly in the freelance era, casting directors were expected "to read a script with the insight required to understand how it would translate to the screen and with the knowledge of acting talent needed to cast every role."[215] It was no longer useful to bring in twenty similar people, but rather, to consider different ways roles could be played and bring the producer or director "five different but very good actors for consideration for a single role, not 100 clones," so that they could cast one of them or at least learn more about the parts.[216]

Although it was more creative, this reformulation of casting also shifted to include more aspects of feminized labor and creative service. The process of reading actors for roles became more intense and multistaged, in turn generating even more paperwork, more phone calls, more names to check, and so forth. With stars now a key part of the package that could greenlight a film—and demanding budget-breaking salaries and perks—freelance casting involved more negotiation among more players on all sides; consequently it was more fraught, dramatic, and emotional. In line with the trend toward flexible specialization, casting directors also let individual TV and filmmaker preferences dictate their work environment, style, and pace on a project-by-project basis, adjusting their process to fit different directors and producers.

Women entered the profession in increasing numbers, and, as it swapped its executive identity for one more akin to a production crew position, assumed many of the feminized duties previously carried out by female talent and clerical workers. Women were so accepted in the field by the late 1970s that, when studios and networks began gender integration of their executive ranks to avoid equal rights–related public scrutiny, they named women vice presidents in casting, rather than production, because "casting was one area of the industry where companies thought it was safe to put women, and where they thrived."[217] By the 1980s, women already dominated the field, and by 2001 their dominance had reached new levels: women received twenty-three out of twenty-four Casting Society Award nominations for feature film work that year, and forty-three of fifty-four primetime TV nominations.[218]

Contemporary Casting: A Hybrid

Some of the traits of contemporary casting do not dovetail neatly with traditional notions of management or with feminization, but they may be viewed as a hybrid of the two, their efficacy representing innovation. One such trait is collaborative leadership, which characterizes many freelance casting companies, in which casting directors work in partnerships of two to four. The arrangement allows more work to be taken in by the collective because, "if you have two partners, you can have eight jobs, whereas, if you just work alone, it's very hard to split yourself up, and you only can do one, maybe two jobs."[219] These partnerships mean that leadership tends to be more collaborative as well. To keep pace with the complexity of multiple projects in multiple phases—each requiring a heavy volume of clerical, informational, and communication work—duties are shared among staffers in a more lateral

network, rather than a single linear chain of command running from casting director to intern.

Though partnerships have long been the chief organizational model of
small businesses trafficking in elite client services (for example, law firms
or ad agencies), casting's level of cooperativity is nonetheless unusual. It
resembles the organizational model of the democratic workplace, in which
management is dispersed more evenly among workers. This model both pre-
dates the twentieth-century de facto system of command-and-control, and
is being reenlisted by organizational experts in line with the new emphases
in that field, which are on group cohesion rather than corporate strategy.[220]
Collaborative leadership is an organizational model identified with female
leaders because, as psychologists and linguists have argued, women are
socialized from childhood in same-sex peer groups to form and maintain
lateral interpersonal networks by seeking consensus, connection, and rapport with others, whereas men, driven toward contest by socialization, seek
status in a hierarchy among opponents and even peers.[221] A more democratic,
collaborative organizational model very well may have found functionality
in the world of casting because it is now largely a female world, in which just
such leadership strategies have had space to take shape.

Casting directors' adaptability, forged through their unusual infiltration
of the field from feminized supporting roles, may also be seen as a useful
innovation for the contemporary industry in general, helping it adjust to
the constant change on the managerial, creative, informational, and emotional levels that characterizes modern production. Flexibility on multiple
levels was an oft-cited characteristic of a skilled casting director, who "helps
from the moment that you do the list or pre-read that actor, pick that actor
and navigate the whole process with them," in a process that's continually
evolving.[222] In a global economy in which technological and economic
complexities demand increasingly flexible, specialized, responsive workers,
casting directors, as adaptive feminized-managerial hybrids, may very well
represent the ultimate service professional.

However, the modern feminine incarnation of casting must not be presented in an entirely positive and progressive light; women, in adapting to
the needs of the client, the marketplace, the industry, and society, have, as
workers, always engaged in acts of professional contortion. Like the traditional wife to whom several interviewees compared themselves, the casting director's ability to perceive and respond to the thoughts and feelings of
others—rather than her own—enables her to affect a film or television text.
Casting directors emphasize the director as the leader of the process, and
the person whose vision they follow, because though "our job might be to
keep everybody happy, it really is to direct the casting process for the director."[223] Descriptions of the ideal relationship between a casting director and

a director often recall the studio secretary's proxying and emotional matching, emphasizing the importance of being "in sync with the director. . . . It is almost like you are trying to crawl into their brain."[224]

Conversely, the producer or director of a project is able to directly control the experience of his casting director. Said one interviewee:

> When you feel like you're just waitress serving up actors where they don't care about your opinion, you oftentimes don't have a connection to it. . . . There are those who hire casting directors as purely secretarial, you know, "Bring in the actors and we'll pick who we like and we don't care about your opinion." And then there are others who hire you to shape the show. They like [your] aesthetic or they want to be challenged.[225]

The power imbalance between casting directors and the project's ultimate authority is most evident when it comes to final decisions. Casting directors generally refuse to tell an employer directly which decision to make; instead, they "help the people making the final decision to find the best people and [avoid] mistakes that they might make out of emotional reasons or getting frustrated. But if a casting director were to just come out and tell those people what to do, chances are that they wouldn't do it."[226]

As such, this creatively vital, high-status version of casting, as reshaped and expanded by female casting directors, continues to resemble the creative service of other, earlier feminized production roles. Though they are creative and certainly demonstrate artistry in their work, casting directors nonetheless exercise these talents under the guise of facilitating the creative vision of a superior. One casting director's statement that casting is "more support art, it's more of a craft in some ways . . . and it's a craft that supports the art of other people" evinces the link to creative service quite unequivocally, whereas others reference this alliance indirectly in statements about getting inside the heads of directors to serve their creative vision. Still others compare casting to the female-associated field of editing.[227] Again, creativity comes through delimitation, as "You're not bringing in the whole world, you're editing the process" by cutting out "90% of what is sent to you and showing the 10% that they need to see."[228]

Combining the obliqueness of the casting director's agency with its loss of studio-era executive status adds up to a loss in professional creative status. Many feel a lack of respect from "an industry that sees us as technicians rather than artists," in which "casting isn't seen as an autonomous creative process."[229] And, like other fields whose feminization undermined them by suggesting they were less rational and involved fewer hard technical and professional skills, the work of casting directors rarely receives due credit.

Casting directors believe in their own creative input to the process as "the person who helps to assemble those actors" in an ultimately successful creative venture by believing in them "enough to present them to our director for a final choice."[230] But, however essential they are to the success of movies and series, casting directors only became eligible for the Emmy in 1989 and are still not eligible for Academy Awards. Ultimately, their work is assessed in terms of negative credit, as blame for poor performances is placed on poor casting, while accolades for stellar casts are bestowed by directors, producers, and actors from awards podiums.

This lack of professional respect is also reflected in the compensation casting directors receive.[231] Marion Dougherty reportedly said that only women could afford to do casting, which makes sense—given that, until 2005, the field had no union to collectively bargain for the salary guarantees, health benefits, and pensions enjoyed by nearly every member of film production crews above the level of production assistant.[232] Tracy Lillienfield explained the delayed unionization as a result of the freelance path that casting had taken, explaining, "We were fighting for jobs and struggling to get people to recognize that there was even a job called 'casting director.' We just worked so hard to do what we do that we forgot to take care of ourselves. I've worked for 23 years and I have no pension."[233]

In truth, casting directors did not simply forget to take care of themselves, but rather they actively chose not to unionize as one of the few strategies available for making creative inroads. In the late 1970s and early 1980s, a group led by Mike Fenton attempted to unionize under IATSE as the Casting Directors Guild Local 726.[234] The faction sought a higher minimum weekly rate, as well as medical benefits, but its efforts were quickly scuttled by another group casting directors—fifty out of sixty-four of them women—who took out an ad in *Daily Variety* stating that its signatories believed that such a union would have a detrimental effect on their casting functions, as well as "the producers, directors and writers with whom we work."[235] The anti-union faction feared that if they submitted to collective bargaining, they would make themselves the enemies of the creative elite, whose recognition and acceptance they had been courting since they became independents. Though unionization was eventually achieved, that casting directors' pay lagged for so long behind professions below them in the production hierarchy affected their overall earning potential.[236]

Conclusion

Casting presents an example of both the far reaches of sex segregation practices begun a century ago, and their affects—both positive and negative—on the contemporary media industry, its products, and practices. Like development workers and continuity workers, casting directors offer ample evidence of just how much the industry's history continues to affect its present-day labor force in which, to quote Miranda Banks, "gender plays into the collaborative nature of film and media production—not just in what is produced but in how." Banks continues, "In subtle ways, much of the work women do in Hollywood is—both through language and through economics—treated as 'women's work.'"[237] Despite efforts to keep their heads down, focus on increasing their field's prestige, and wait for their creative contributions to be recognized by the creative elite and production crew with whom they work, casting directors have found it difficult to lay down the gendered baggage of the past. For, although casting didn't truly emerge as a woman's profession until the 1970s, the conditions for the shift were established half a century earlier with the demarcation of certain kinds of film labor as "women's work."

The logic that has endorsed gendered labor since the turn of the twentieth century—and been sustained by industry mythology—has determined that, whether in 1910 or 1970, when media work aligned with a feminized labor sector and shed managerial/executive identity, it fell to women who, in times of change, could be counted on to do more for less, and to absorb post-studio costs through their freelance labor. As in other film-specific "women's" fields, casting directors were able to acquire additional creative capital and to innovate in leadership and organizational style. But the price for these gains has been high, demanding inconspicuousness in terms of creative agency and credit, requiring careful management of status and gender, in exchange for a work experience shaped largely by the preferences of others.

Women's positive impact on the industry from feminized fields is obvious in the ways casting has evolved in the hands of a predominantly female workforce that has pushed the field forward in terms of artistry. However, a gender binary persists; when women are attributed skill in one area, the implication is that either they are worse in others, or that that area somehow demands lesser skill or holds less importance in the first place. Yet it is some consolation that, contrary to media history narratives about women's near–total exclusion from film production after the 1910s, women have found ways to participate in creative media production at high levels and to make their mark on media history by turning to their advantage precisely the nonthreatening "feminine" skills that have been used to disqualify them and relegate them to typewriters and sewing machines.

Epilogue

The Legacy of Women's Work
in Contemporary Hollywood

> He said that, as a woman, my best bet to
> land a writer's assistant job was to babysit for a
> [male] writer first. . . . Tell me how the guys get in
> and how 'bout I try that?
>
> —Anonymous poster, "Shit People Say
> to Women Directors . . ." (2015)

Blogger Nikki Finke's 2009 report on the pay cut planned for assistants at the newly merged William Morris Endeavor Entertainment (WME) talent agency was met with a surprisingly vicious response from her readers.[1] The responses, which appeared on the comment thread of her *Deadline Hollywood Daily* (DHD) blog, were not surprising simply because they were angry—indeed, some venting was to be expected by and on behalf of the assistants in question, given that what was at stake was a nearly 20 percent drop in salary (from $13.50/hour to $11.00/hour). Rather, the anger of the commenters was surprising because it was expressed most frequently and most vehemently in *support* of management. Commenters applauded the architects of the pay cut while admonishing assistants that they were "not coal miners," and telling them to "sack the fuck up and deal with it," to "grow up!!" because "this is the business, always has been and always will be—it's called paying your dues by working long hours, taking it up the ass, and not getting paid."[2] These respondents further insisted that the assistants would be foolish to strike, since "in this town, walkout equals lock out," and they would quickly be replaced by others "chomping at the bit to take their place."[3]

Though Finke's blog was de rigueur reading for many Hollywood insiders at the time it was written, it was also unaffiliated with a studio, agency, or network. Therefore, it might be easy to dismiss the comment dustup as a one-off Internet flame war, separate from the entertainment industry proper and its system of production. In reality, the blog and its comments are very much part of the industry. In fact, they represent a perfect example of its most important sustaining function: the development of industrial mythology that, over time, is codified and dispersed as fact through its creative products, and by and to its members. It is this mythology that surrounds and supports otherwise chaotic post-Fordist twenty-first-century system of media production, helping to contain and control the contributions of low-status workers—like the assistants in question—to the work of writers, producers, directors, agents, and executives. Such mythology (evinced by claims to the contrary by the industrial boundary police on Finke's blog and elsewhere) disguises the fact that assistants are absolutely essential to the industry, and do affect it creatively, just as they did fifty years ago when they were more commonly labeled secretaries and their role was understood as women's work.

Absent the rigid studio system of the classical Hollywood era, with its clear physical and organizational boundaries, it is through Hollywood's production culture and the texts it produces that today's workers and aspirants "make sense" of the entertainment industry, when in reality, as John Caldwell points out, "production cultures are far too messy, vast and contested to provide a unified code . . . for breaching [their] walls."[4] Caldwell argues that industrial mythology—the way media work is rationalized and framed through its workers' trade stories—helps to keep an increasingly divergent mixture of industrial modes from diverging completely.[5] This mythology is especially useful in keeping the work of assistants to writers, producers, directors, agents, and executives low status and cheaply compensated. Having first worked as an assistant and later as a researcher conducting ethnographic interviews with other assistants, I designate the advice, stories, and "industrial wisdom" surrounding the concept of "dues paying"—the process of "paying dues" at low-status job in order to prove worthy of a higher-status position—as the sector's overarching mythology. On the Finke blog, the need for assistants and other entry-level workers to pay their dues is the only thing that all commenters agree on. Many times throughout the blog's 172 comments, assistants and their defenders qualify remarks by insisting that they know they must pay dues, while the pro–pay-cut faction accuse assistants of not being willing to go far enough, insisting they accept their fates, because, as one commenter admonished, "This is the business, always has been and always will be."[6] Statements like these show just how effectively dues-paying mythology reins in worker dissent since, contrary to the industrial wisdom they purport to impart, the dues-paying model of suffering-for-entrée is by no means the way it "always has been." In fact, for the first sixty-plus years of the entertainment industry, this model didn't exist. As these chapters have demonstrated, what did exist was a sector of the workforce that was allotted low pay and status because it was feminized.

Like the studio secretary's labor, assistant work is usually some combination of clerical and administrative work, personal errands, and creative-executive–level work on an employer's projects, as well as a heaping helping of service and emotional labor. Much like secretaries at MGM or Warner Bros. in the 1930s, today, some assistants report having supportive, mentoring employers and others wind up working the desks of bosses who are abusive and who assign them only unrewarding tasks.[7] Despite this range of experiences, the mythology that has evolved around dues paying tends to elevate and celebrate the harshest, most embarrassing extremes of assistanthood and the most abusive "screamer" bosses. These sorts of grotesque, hero's-journey narratives explain the media industry as a world that can be navigated and mastered by those willing to accept its rules. Instead of improving upon the studio secretary's lot, under this mythology, today's assistant is just as helpless to avoid abuse. The only real difference between then and now is that whereas studio secretaries' were expected to suffer for their bosses simply because they were women (service being their gender's specialty and a reward unto itself), for today's assistant, doing anything and everything requested by an employer, no matter how strange or difficult, is framed as the only way to really "make it" and be promoted through the ranks. From TV shows to chick-lit novels to networking websites, stories of assistants completing unsavory tasks and taking vile abuse circulate endlessly. The stories might be mistakable for simple venting if they didn't so often carry the moral that these sorts of experiences should be not shunned, but rather desired as the trial-by-fire necessary for advancement. Just one example of that moral at work can be found in the how-to manual *It's All Your Fault! (How to Make It as an Assistant in Hollywood)* that describes assistanting as "a purgatory" and "a rite of passage which, if suffered and surmounted, can land you in those heavenly regions beyond the pearly gates of Brentwood or Los Feliz."[8]

Though the ranks of assistants include far more men than the secretarial pools at studios, assistanthood is still associated with women more than men, and assistants still do women's work, whether or not the work is framed as such. Though the job has remained the same, the mythology has been reformulated to reflect the fact that men also hold assistant roles. Episodes of HBO's behind-the-scenes comedy *Entourage* are peppered with examples of assistants who find themselves in strange, often uncomfortable positions and succeed by "working" those positions (that is, doing whatever it takes to ensure the success of their boss or his movie, and by extension, themselves). For *Entourage's* female assistants, "working it" in service of an employer often involves having sex for or with them. In an episode entitled "Sorry, Ari," an agent attempting to woo actor Vincent Chase to his agency tells members of his entourage that they won't have to worry about being rejected by the women who work in his office because "all my assistants love to fuck."[9] Throughout the series, sexual willingness on the part of female underlings is portrayed as a good thing, the kind of team spirit that will lead to a bright future. When his gay male assistant is similarly propositioned in exchange for benefits to his employer, power agent Ari Gold intercedes at the last minute to save him from the same sort of sexual exploitation that Gold engages in with female workers on a regular basis. The message that female assistants should be good sports to prove their worthiness of membership in elite media circles is reinforced repeatedly via the character of Dana Gordon, a female studio executive who, according to Gold, served as his assistant and sex partner before her big break—a fact that he still uses to extract favors and confidences from her years later. This isn't to say that nonfictional assistants really do consider sex as a part of their job descriptions. In fact, most do not. Instead, more disturbingly, these kinds of portrayals imply that, according to the media makers who created them, every female assistant *should*, hinting at the ways in which women's work is valued differently than men's and how gendered expectations still shape women's experience of media production.

In studio days, secretaries and female executive assistants were compensated unfairly and victimized by male employers because they were women. Today, the same conditions have been allowed to continue because of the benefits they offer management and have been grandmothered in under the cover of paying dues and proving worthy of promotion.[10] Boundaries between reasonable and unreasonable work, hours, pay, and so on blur, as unpleasant and unjust aspects of assistant work are framed as merit badges that might qualify the assistant for inclusion the ranks of media elite rather than as what they really are: the same exploitation of labor that originated under rules of gender-based exclusion. Additionally, despite great social change since the studio era, many socially constructed, essentialist notions of men's versus women's skills remain in the minds of both workers and employers, even though language in job ads and corporate documents is coded to appear gender neutral. These lingering effects of earlier feminization help explain the continued low status and compensation of assistants. The shift of assistant identity from one of feminized ghetto to stepping stone explains the rest.

At the same time that secretarial labor for movie makers began to be gender integrated in the late 1970s and early 1980s, it was reframed discursively, shifting it in workers' minds from a feminized sector that was compensated as such (set apart from men's fields in terms of pay, credit, and promotion), to a dues-paying, quasi-apprenticeship, in which the pay was low because the true compensation was not money, but training for their future profession under one of its masters. Though it goes by various other names (trainee programs, internships, assistant hyphenates), apprenticeship is common practice in unionized, below-the-line film professions, so it makes sense that it has been adopted by above-the-line professionals and agents. However, the difference between an assistant location manager and a producer's assistant is that, as with any real apprenticeship, the assistant location manager's union—the 399 Teamsters—regulates the length, pay, and conditions of their apprenticeship and determines when they are eligible for promotion and pay raise. By contrast, there are no rules governing assistants' pay, how they earn it, and what kind of training or promotions they may receive.

Really, assistantship hasn't changed from the days of secretarial work as much as it has been rhetorically rebranded to have the appeal of an apprenticeship without any of its guarantees. Hence, the widely varying pay and work experiences of contemporary assistants. The mythology of dues-paying retrofits the parts of the apprentice system that are most favorable to management onto formerly gendered aspects of the job that are equally favorable, then spins the new combination as not only attractive but a key to success in the business. The sanctioning of work conditions through dues-paying might be more defensible if the promises of the dues-paying mythology were more consistently delivered. However, far from assistants finding themselves on track to creative or executive positions after a few years on the job, there is an extremely large dropout rate due to the financial, mental, and emotional strain endured by workers for up to ten years without promotion.[11] Pay has remained low and unregulated (between $500 and $900 per week in a city where the average rent is roughly $1,600 a month) because even though the work is no longer reserved for women, assistants are led to believe that the real compensation is not money but opportunity for advancement.

Dues-paying mythology makes little mention of the contributions assistants make to the creative process. However, like the studio secretaries, assistants find themselves participating in creative projects through a contemporary form of creative service, which includes a similar range of responsibilities, from proxying for and filtering noncreative work away from employers, to delimiting creative solutions to problems (for example, generating writer lists or narrowing a field of a hundred scripts to the best five), to actively generating content (such as by creating transmedia texts). Assistant work's impact on creative texts was not as clearly delineated as that of those higher in the creative food chain, who like their movie-maker forerunners typically characterize assistanting as ancillary or parallel to the creative process. Yet as members of the network that creates screen content, assistants are clearly essential to the process, responsible for a large portion of the distributed knowledge that goes into collaborative efforts, and quietly carrying out a large share of the development and production labor, especially that which takes place on paper. What is more, in the past decade a growing number of firms have eliminated many of their mid-level positions (for example, creative executive) and simply distributed their duties to an expanded class of cheaply compensated assistants and interns, who compete for an even smaller number of promotions out of that sector. The position of story editor, which for Samuel Marx at MGM was an important management role, is understood at most contemporary production companies as an assistant job with additional executive responsibilities. In addition to processing all incoming story material and managing the company's freelance readers, story editors typically "pick up phones" (read: serve as assistant) for one or more executives. The salary for doing the work of both a junior executive and an assistant is roughly $30,000–$35,000 per year or $600–$700 per week (an assistant's pay). The majority of workers who accept these jobs are women, many of them working seventy-hour weeks for years before being promoted or leaving the job.

Far from being ignorant of the discrepancy between their value to the system and their compensation, most assistants recognize it and accept it, at least early on in their tenures. California labor laws for overtime are broken every day in Los Angeles and, until the past few years, offenses were seldom reported, and attempts at unionization were rare and fruitless because under the rules of dues paying, complainers would "never work in this town again." This mythology has been the focus of greater scrutiny only recently, since a judge ruled in favor of plaintiffs in a 2013 lawsuit brought by Fox Searchlight interns claiming illegal unpaid labor in the production of *Black Swan* (2010), prompting speculation about further legal action by unpaid workers and unionization or other measures by their underpaid assistant brethren. Even so, most assistants continue to accept their fate because most of them believe that if they pay their dues, they will be rewarded.[12] And when they aren't, the mythology is in place to shift the blame from management to the assistants for being unwilling to pay their dues. Of course, the only definite beneficiary of this policy is the industry that continues to sustain itself on these workers' backs. This helps to explain DHD blog commenters with seemingly little stake in the WME wage dispute angrily championing management over labor: whether they have transcended assistanthood, are themselves assistants, or are outsiders looking in, the commenters do have a stake in the argument—a belief in the system as rational and in their own ability to succeed if only they can master it. And, as they have been mentored by television or job gossip to do, they defend the system, policing its boundaries and its mythology. This rationalization sanctions the system, helping it contain exactly the kind of dissent that, ironically, might ultimately make it fairer and more masterable.

I offer this example of assistants and dues paying, as well as that of the current-day casting directors in the previous chapter, to provide a sense of the complex ways in which the gendered past continues to reverberate for present-day media workers and is reproduced in the products their industry produces. Today, gender-related expectations still exist, but they're difficult to isolate because they have been forced under the radar in order to give the appearance of political correctness. My own experiences in media production reflect the ripple effect of historical feminization in similarly murky ways. Early on, in my internships at various production companies, I felt myself being encouraged toward work as an assistant or in development, offered the rationale that women excelled there, if not in upper-level production roles. A frequent source of annoyance among the female assistants at my first full-time job was the fact that it was always one of us (rather than the male assistants) who was assigned the task of placing large orders for catered lunches and setting them up in the company's conference room. At another office, my female employer—whose films about women had marked her publicly as feminist—told me I should hire a male production assistant because the PA job occasionally included some lifting and carrying and "women can't carry things." And always there was the sense that, though many production companies employed female executives, they never outnumbered men, and that this was the natural order of things.

These were minor gripes, but they continued to accumulate over the next four years. What bothered me most was that I could not talk about any them. At least, not out loud in the office. In an industry where nondisclosure agreements were standard, non-union assistant gigs were the only way in to elite creative fields, and women were still frequently framed as interlopers whose responsibility it was to prove they could fit into a workplace implicitly understood as belonging to men, keeping quiet seemed like the best way to be kept around. I've kept in contact with the female assistants I "came up" with, as well as students who have since entered the industry and many of the other female movie workers I've come to know through my interviews and in my work and social life in Los Angeles. For those women still working in media production, the anecdotes relating to their gender seem to accumulate, their effects magnified as they rise up the ladder. The TV writer notices that the producers who hire her invariably assume that she is most interested in writing stories around female characters and themes simply because she's a "lady writer." The female second AD feels compelled to tell "dick jokes" on the set in order to prove she "gets it" and can hang with the guys. The writer's assistant is teasingly blamed by coworkers for a mistake, prompting the office's male executive producer to joke that it's okay because "when she makes enough mistakes, we get to rape her." Though everyone around him knows the comment is shockingly inappropriate, no one reports it for fear of rocking the boat, and for years, every time one of her former boss's projects is announced in the trades, the former assistant feels like she belongs in her industry a little less. I'm not the only one hearing these stories. When the Tumblr blog "Shit People Say to Women Directors (and Other Women in Film)" appeared in the spring of 2015, it quickly accumulated hundreds of stories related by anonymous and not-so-anonymous posters identifying as women working in film and television, recounting how they'd been told to wear more makeup, that their technological skill was surprising because they were women, that they'd never be able to get a crew to follow their instructions because they were women, and so on.

And still, discussion of women in media production, whether by media workers, journalists, or in media texts, centers around phrases like "breaking in," "infiltrating," "storming the castle," and so forth. Discursively, women continue to be attributed less ownership over their industry and its history than men, as if they really were absent from production prior to the 1980s. The truth is that, unlike older industries that were exclusive to men for decades or centuries before women sneaked in as clerical workers, women have been present in and around media production all along, often subsidizing growth through their labor's cheaper cost; and that women aren't born better at script supervision or publicity any more than they are born with an innate sense for how much food to order for lunch or for noticing details. They've excelled in those areas because those were the areas in which they were allowed and encouraged to excel.

Scholarship and pedagogy can play a larger role in shifting entrenched thinking around marginalized groups and their place in media history as well as the contemporary media industry. A partial corrective lies in teaching students who plan to pursue careers in media or creative industries how constructed notions of gender (and race, class, and so on) continue to affect media labor, as many media studies programs do. However, more may be done to draw attention to the ways in which these conceptions of gender affect how creative labor is distributed and compensated, how creative industries often frame attempts social or political change—like hiring more women—as tradeoffs for creative goals or artistic freedom, and how all of this affects the end product. This sort of training and preparation, if undertaken prior to joining the industry, might help newcomers recognize the real reasons women aren't streaming into the Directors Guild and resist conventional wisdom about how women's inherent unsuitability.

Proper, critical unpacking of these issues and their origins in film history might help the next generation of workers discard outdated mythology and ask, for example, whether a joke is inherently funny to men and non-buzz-kill women alike, or whether it's funny because men have been the only ones with any say in what's funny for too long. It might ensure that female writers' "good female characters" are recognized as the result of their individual talents, not their ownership of a set of female sex organs. Providing historical context in the classroom might guide future employers toward hiring workers based on their difference from everyone else at the table, rather than their sameness . . . not because that's the Right Thing to Do, but because, when no one type of person is more "naturally" suited for a particular kind of work than any other, it makes good business sense to seek a variety of creative viewpoints.

Equipping future female media workers with a sense of legacy might help them seek out the sector that's best for them based on their own desires, not their understanding of what's best given their gender. And for those women already in them, a better understanding of women's sectors might engender a new sense of ownership over women's work. Instead of agreeing with negative characterizations of this work as unimportant "scut" or "bitch" work, workers might feel an owner's pride, demanding appropriate compensation, or else leave such sectors behind more readily. I disliked assistant work intensely. I still enjoy my work as a reader of books for L.A.-based production companies. I feel proud of my work in both sectors, in part because I connect it to the work of Marcella Rabwin or Kate Corbaley and recognize both the positive and negative sides of its historical lineage. If more women could do the same, they might more readily demand that both industry and academy recognize their work and feel entitled to advance into those fields that have been dominated by men, not as an unlikely act of heroism, but as their rightful place in their business after over a century of contributions.

APPENDIX
WORK ROLES DIVIDED BY GENDER AS REPRESENTED IN STUDIO TOURS FILMS

Behind the Screen (Universal, 1915)

Men's Labor
Actor (15–20)[1]
Producer/Production Head (1)
Costume Supervisor (1)
Director (2–4)[2]
Lighting Department (1)
Property Department (20–30)
Scenario Department[3] (4)
Scenery Department (3)
Scenery Supervisor (1)
Set Dressers/Grips (15–20)
Waiter (4–5)

Women's Labor
Actor (5–10)
Note-Taker/Possible Script Clerk (1)
Scenario Department (1)
Seamstress (7)

A Trip to Paramountown (Paramount, 1922)

Men's Labor
Actor (15–20)
Camera (10–12)
Director (4)
Set Painter (1)

Women's Labor
Actress (8–10)
Writer (1)
Maid/Servant (1)

A Tour of the Thomas H. Ince Studio, 1920–1922 (Ince, 1924)

Men's Labor
Actors (25+)
Butler (1)
Camera Department (10+)
Cameraman (6+)

Women's Labor
Actors (10–15)
"Manikins"/Models (4)
Negative Inspector (9)
Note-Taker (1)

A Tour of the Thomas H. Ince Studio,
1920–1922 (Ince, 1924) *(continued)*

Men's Labor
Carpenter (12+)
Chauffer (1)
Director (4–6)
Editor (1)
Firefighters (3–4 plus several volunteers)
Grip (3)
Messenger/Porter (1)
Negative Developing (4)
Negative Drying Room (6–8)
Negative Timing Technician (2)
Plaster Molder (3)
Power Plant (3)
Powder/Explosives (1)
Producer (2–3, including Ince)
Production Designer (3)
"Production Staff" (7)
Projectionist (1)
Professional Fitness Trainer (1)
Set Designers (5)
Scenic Artist (3)
Set Decorator (5)
Still Photography Department (3)
Studio Manager (1, Ince)
Titling Artist (9)
Violinist (2)
Waiter (1)

Women's Labor
Patcher (9)
Pianist (1)
Seamstress (2)
Wardrobe Department (3)
Wardrobe Maid (1)

1925 Studio Tour (MGM, 1925)

Men's Labor
Actor (10–15)
Art Director (1)
Art Director's Aide (10–12)
Art Studio (4–5)
Assembly Supervisor (1)
Barber Shop (5)
Boss Painter and Decorator (1)
Busboy (3)
Camera Repair Shop (3–4)

Women's Labor
Actor (10–15)
Art Director's Aide (1)
Assembler (7–8)
Barber Shop (1)
Cutter (2)
Dancer (15)
Danseuse (1)
Laboratory (10–15)
Matron (1)

Carpenter Shop (20–30)
Casting Department (3)
Casting Director (1)
Construction Manager (1)
Costume Designer (1)
Cutter (18)
Director (18–20)
Draftsmen (5)
Electricians (20–30)
Laboratory (25–30)
Laboratory Head (1)
Lighting Supervisor (1)
Orchestra Units (18–20)
Painter (30–40)
Pianists (5–6)
Plastic Shop (10–15)
Production Manager (1)
Production Manager's Staff (17–18)
Projectionist (12+)
Props (15–30)
Publicity Department (10)
Reader (3)
Reception Clerk (5)
Screenwriter (5)
Scenario Writer (10)
Set Builder (50–60)
Still Cameraman (15–20)
Transportation Department (1)
Wardrobe Department (2)
Wire Operator (1)

Nurse (1)
Pianist (1)
Publicity (1–2)
Reader (5)
Research Department (2)
Scenario Writer (8)
Screenwriter (4)
Seamstress (??)
Stenographer/Typist
Violinist (1)
Wardrobe Mistress/Matron (1)
Wardrobe Staff (?)

Universal Studio and Stars (Universal, 1925)

Men's Labor
Actor (30–40)
Advertising Manager (1)
Camera Department (5–10)
Director (5–10)
Ranch Rider (10–20)
Reception Clerk (1)
Scenario Editor (1)
Scenario Writer (1)

Women's Labor
Actor (15–20)

Life in Hollywood (Goodwill, 1927, showing labor at Vitagraph)

Men's Labor
Actor (10–12)
Cameraman (3)
Clapper Loader (1)
Director (8–10)
Violinist (1)

Women's Labor
Actor (8–10)
Fan Mail Clerk (1)

A Trip thru a Hollywood Studio (Vitagraph, 1935)

Men's Labor
Actor (4)
Boom Operator (1)
Cameraman (2)
Camera Department (5)
Casting Assistant (1)
Casting Director (1)
Cutter (1)
Dance Director (1)
Director (2)
Executives (4)
Laboratory Drying/Transfer Rooms (1)
Laboratory Printing/Matching Room (5)
Lighting Department (3)
Pianist (1)
Power Plant Operator (1)
Projectionist (1)
Sound Amplifier Operator (2)
Sound Mixer (1)
Sound Recorder (1)
Still Photographer (1)
Trainer (1)

Women's Labor
Actor (15–20)
Continuity/Script Supervision (1)
Dancer (25–30)
Negative Cutter (7)
Note-Taker/Secretary (1)

20th Century Fox Studio Tour (1936)

Men's Labor
Actor (15–20)
Arsenal (2)
Art Director (1)
Asst. to Makeup Department Head (1)

Women's Labor
Actor (15–20)
Costume Design/Dept. Head (1)
Dancer (15–20)
Film Editorial Department (1)

Associate Producer (9)
Camera Operators (3–4)
Camera Repair (7–8)
Camera Repair Head (1)
Carpenter (2)
Casting Director (2)
Chief Engineer (2)
Composer (1)
Construction Department Head (1)
Costume Design/Dept. Head (3)
Cutter (1)
Dance Director (1)
Director (11–12)
Draftsmen (15–20)
Eastern Scenario and Play Editor (2)
Exec. Assistant to Production V.P. (1)
Executive Producer (2)
Film Editorial Assistant (1)
Film Editorial Department Head (1)
Film Laboratory (4)
Fire Chief (1)
Lumber Mill (5–6)
Makeup Department Head (1)
Maintenance Department (3–4)
Maintenance Department Head (1)
Miniature Processing Department Head (1)
Music Associates (5)
Music Department Directors (3)
Payroll Clerk (1)
Police/Armed Guards/Firemen (50+)
Police Chief (1)
Power Generator Operator (1)
Personnel Manager (1)
Production Manager (1)
Property Department Head (1)
Public Relations Head (1)
Scenario Department Head (1)
Scenic Art Department Head (1)
Scenic Art Crew (5)
Screenwriter (12)
Set Designers (3)
Sound Assistant (2)

Research (5)
Research Department Head (1)
Screenwriter (4)
Wardrobe Department (15)

20th Century Fox Studio Tour (1936) *(continued)*

Men's Labor
Sound Department Head (1)
Sound Technician (1)
Story Editor (1)
Studio Manager (1)
Studio Treasurer (1)
Transportation (3–4)
Vice President—Production (1)

NOTES

Introduction

1 Edison colorists: Charles Musser, *Before the Nickelodeon: Edwin S. Porter and the Edison Manufacturing Company* (Berkeley and Los Angeles: University of California Press, 1991), 41. "American Mutoscope: The Art of Moving Picture Photography," *Scientific American*, April 17, 1897, 248–250; "deft-fingered girls": "Selig's The Great Moving Picture Plant of the West," *Moving Picture World* (hereafter referred to as *MPW*) 5, no. 8 (August 21, 1909): 248; "Girl operator": Melvin Riddle, "From Pen to Silversheet XIV—The Film Laboratory," *Photodramatist* 4, no. 9 (February 1923): 6, 38. "Sample copy room": Edward Dmytryk, *It's a Hell of a Life but Not a Bad Living: A Hollywood Memoir* (New York: Times Books, 1978), 5.

2 Greer photograph: Christopher Finch and Linda Rosenkrantz, *Gone Hollywood: The Movie Colony in the Golden Age* (New York: Doubleday, 1979), 69. "Embroidery": Ronald L. Davis, *The Glamour Factory: Inside Hollywood's Big Studio System* (Dallas, TX: Southern Methodist University Press, 1993), 216; Beth Day, *This Was Hollywood: An Affectionate History of Filmland's Golden Years* (New York: Doubleday and Company, 1960), 128–129. Disney/"the nunnery": Patricia Zohn, "Coloring the Kingdom," *Vanity Fair* (March 2010) http://www.vanityfair.com/culture/features/2010/03/disney-animation-girls-201003.

3 Z. Richter, "Feminization of Labor," *Wiley-Blackwell Encyclopedia of Globalization* (London: Blackwell, 2012). http://onlinelibrary.wiley.com/doi/10.1002/9780470670590.wbeog201/full.

4 Organisation for Economic Co-operation and Development, *The Future of Female-Dominated Occupations* (Paris: Author, 1998), 9–10.

5 Arlie Russell Hochschild, *The Managed Heart: Commercialization of Human Feeling* (1983; reprint with new afterword, Berkeley: University of California Press, 2003), 7.

6 Vicki Mayer, *Below the Line: Producers and Production Studies in the New Television Economy* (Durham, NC: Duke University Press, 2011), 133–134.

7 Gender performativity introduced and elucidated in Judith Butler, *Gender Trouble* (1990; reprint, London: Routledge, 1999).

8 Jane Gaines, "Film History and the Two Presents of Feminist Film Theory," *Cinema Journal* 44, no. 1 (Fall 2004): 117.

9 Invisibility and erasure of women's labor in contemporary media production examined in Miranda Banks, "Bodies of Work: Rituals of Doubling and the Erasure of Film/TV Production Labor" (PhD diss., University of California, Los Angeles, 2006); Mayer, *Below the Line*.

10 The line discussed: "There is actually a heavy black line on the budget sheet which separates these two types of costs." Mike Steen, *Hollywood Speaks* (New York: G. P. Putnam's Sons, 1974), 187.

11 Leo Rosten, *The Movie Colony, the Movie Makers* (New York: Arno Press, 1941), 32.

12 Ibid.

13 From studies of women as filmmakers and audiences such as Anthony Slide's *Early Women Directors* (New York: A. S. Barnes, 1997) and *The Silent Feminists* (London: Scarecrow, 1996), and Shelley Stamp's *Movie Struck Girls: Women and Motion Picture Culture after the Nickelodeon* (Princeton, NJ: Princeton University Press, 2000) to more comprehensive histories and collections encompassing these eras, such as Musser *Before the Nickelodeon*; Richard Koszarski's *An Evening's Entertainment: The Age of the Silent Feature Picture, 1915–1928* (Berkeley: University of California Press, 1990); Janet Staiger's work on modes of production in *The Classical Hollywood Cinema: Film Style and Mode of Production to 1960* (New York: Columbia University Press, 1985); John Lewis and Eric Smoodin, eds., *Looking Past the Screen: Case Studies in American Film History* (Durham, NC: Duke University Press, 2007); and Steve Neale, ed., *The Classical Hollywood Reader* (London: Routledge, 2012).

14 Cari Beauchamp, *Without Lying Down: Frances Marion and the Powerful Women of Early Hollywood* (Berkeley: University of California Press, 1997); Jane Gaines, "Pink-Slipped: What Happened to Women in the Silent Film Industry?" *Blackwell's History of American Film*, ed. Roy-Michael Grundmann, Art Simon, and Cynthia A. Lucia (London: Blackwell, 2012), 155–177; Shelley Stamp, *Lois Weber in Early Hollywood* (Berkeley: University of California Press, 2015); Kay Armitage, *The Girl from God's Country: Nell Shipman and the Silent Cinema* (Toronto: University of Toronto Press, 2003); Amelie Hastie, *Cupboards of Curiosity: Women, Recollection, and Film History* (Durham, NC: Duke University Press, 2007); Hilary Hallett, *Go West, Young Woman!: The Rise of Hollywood* (Berkeley: University of California Press, 2013); Lizzie Francke, *Script Girls: Women Screenwriters in Hollywood* (London: BFI, 1994). To name just a few more: Jennifer M. Bean and Diane Negra, eds., *A Feminist Reader in Early Cinema* (Durham, NC: Duke University Press, 2002). Antonia Lant, ed., *The Red Velvet Seat: Women's Writing on the First Fifty Years of Cinema* (London: Verso, 2006); Joan Simon, ed., *Alice Guy Blaché, Cinema Pioneer* (New Haven: Yale University Press in association with Whitney Museum of American Art, 2009); Wendy Holliday, "Hollywood and Modern Women: Screenwriting, Work Culture, and Feminism, 1910–1940" (PhD diss., New York University, 1995); *Women Film Pioneers Project*, https://wfpp.cdrs.columbia.edu/.

15 For an example of the breadth and depth this project has taken on, see Jane Gaines and Radha Vatsal, "How Women Worked in the US Silent Film Industry," in Jane Gaines, Radha Vatsal, and Monica Dall'Asta, eds., *Women Film Pioneers*

Project, Center for Digital Research and Scholarship (New York: Columbia University Libraries, 2013). Web, November 18, 2011. https://wfpp.cdrs.columbia .edu/essay/how-women-worked-in-the-us-silent-film-industry.

16 Karen Ward Mahar's *Women Filmmakers in Early Hollywood* (Baltimore: Johns Hopkins University, 2006) examines the initial success of female movie makers in the 1900s and 1910s and their gradual exclusion from most movie maker roles in the late 1910s and 1920s. Mark Garrett Cooper's *Universal Women: Filmmaking and Industrial Change in Hollywood* (Chicago: University of Illinois Press, 2010) focuses on the ranks of female directors at Universal in the 1910s.

17 Soft systems described in Brian Wilson, *Soft Systems Methodology: Conceptual Model Building and Its Contribution* (New York: John Wiley & Sons, 2001), ix; see also Peter M. Senge, *The Fifth Discipline: The Art and Practice of Learning Organization* (New York: Currency Doubleday, 1990).

18 Integrated approaches to the study of people in anthropology outlined in Sherry Ortner, *Making Gender: The Politics and Erotics of Culture* (Boston: Beacon Press, 1996), 1–4.

19 John Caldwell, *Production Culture: Industrial Reflexivity and Critical Practice in Film and Television* (Durham, NC: Duke University Press, 2008); Mayer, *Below the Line*; Felicia Henderson, "The Culture behind Closed Doors: Issues of Gender and Race in the Writers' Room," *Cinema Journal* 50, no. 2 (Winter 2011): 145–152; Miranda Banks, "Gender below the Line: Defining Feminist Production Studies," in *Production Studies: Cultural Studies of Media Industries*, ed. Miranda Banks, John Caldwell, and Vicki Mayer (London: Routledge, 2009), 87–98.

20 Shop Floor Practice in sociology is discussed by Howard Becker in "An Epistemology of Qualitative Research," in *Ethnography and Human Development: Context and Meaning in Social Inquiry*, ed. R. Jessor, A. Colby, and R. A. Shweder (Chicago: University of Chicago Press, 1996), 53–71.

21 One exception to this rule is the University of Southern California Warner Archive, which holds the records from such departments as personnel at that studio.

22 Radha Vatsal, "Reevaluating Footnotes: Women Directors of the Silent Era," in *A Feminist Reader in Early Cinema*, ed. Jennifer M. Bean and Diane Negra (Durham, NC: Duke University Press, 2002), 119.

23 In addition to the many excellent studio histories cited throughout this book, such as Tino Balio's *Grand Design: Hollywood as a Modern Business Enterprise* (Berkeley: University of California Press, 1993), and Richard Jewell's *The Golden Age of Cinema: Hollywood, 1929–45* (London: Wiley Blackwell, 2007), I have found several popular press histories from the 1960s and 1970s particularly useful, in part because to interest their broader audience, they concern themselves with the minutiae that meticulous histories of studio business practice and products often necessarily leave out, such as what the studio commissary was like, which stars had maids. Day, *This Was Hollywood* and Christopher Finch and Linda Rosenkrantz's *Gone Hollywood: The Movie Colony in the Golden Age* (New York: Doubleday, 1979) are cited frequently, often in combination with Davis, *The Glamour Factory* and Steven Bingen, Stephen X. Sylvester, and Michael Troyen, *M-G-M: Hollywood's Greatest Backlot* (Solana Beach, CA: Santa Monica Press, 2011) and other pictorial

histories, which help to corroborate their data with direct quotes and photographic evidence. These popular press histories were less concerned with the *why* of the studio system than with the bits and pieces of studio life that might interest a popular audience, and as such, they managed vacuum up many of the details that escaped notice elsewhere for the very fact of their mundaneness.

24 Denis Wood, *The Power of Maps* (New York: Guilford, 1992), 1.

25 Hastie, *Cupboards of Curiosity*, 6; Walter Benjamin essay in *Reflections: Essays, Aphorisms, Autobiographical Writings*, ed. Peter Demetz (New York: Harcourt Brace Jovanovich, 1978), 3–60.

Chapter 1   Paper Trail: Efficiency, Clerical Labor, and Women in the Early Film Industry

1 Janet Staiger, *The Classical Hollywood Cinema: Film Style and Mode of Production to 1960* (New York: Columbia University Press, 1985). I lay this foundation for later explanations of feminized labor sectors in the film industry through existing secondary resources on women's labor. Most important, and cited on the following pages, Alice Kessler-Harris's *Out to Work: A History of Wage-Earning Women in the United States* (1982; reprint, 20th anniv. ed., Oxford: Oxford University Press, 2003) for discussions of women's introduction to factory and office; Sharon Hartman Strom's *Beyond the Typewriter: Gender, Class and the Origins of Modern American Office Work, 1900–1930* (Chicago: University of Illinois Press, 1992), the reorganization of the office; Marjorie Davies's *Women's Place Is at the Typewriter: Office Work and the Office Worker, 1870–1930* (Philadelphia: Temple University Press; 1982, women's office work; Lisa Fine's *The Souls of the Skyscrapers: Female Clerical Workers in Chicago, 1870–1930* (Philadelphia: Temple University Press, 1990), debates around women's fitness for office work; Leah Price and Pamela Thurschwell's *Literary Secretaries/ Secretarial Culture* (Hampshire, England: Ashgate, 2005) for its cultural studies of both real and fictional clerical workers. Primary resources on this topic come in the form of vocational guides for women such as *Careers for Women* (ed. Filene), Miriam Simons Leuck's *Fields of Work for Women* (New York: D. Appleton and Company, 1926), and secretarial handbooks and how-to manuals cited throughout.

2 For example, see Sam Spade's devoted secretary, Effie Perine, first featured in *The Maltese Falcon*. Dashiell Hammett, *The Maltese Falcon* (New York: Alfred A. Knopf, 1929).

3 Clerk characters appear in at least nine of Charles Dickens's novels, most notably *A Christmas Carol* (Bob Cratchet) and *David Copperfield* (Uriah Heep). Charles Dickens, *A Christmas Carol* (London: Chapman and Hall, 1843); *David Copperfield* (London: Chapman and Hall, 1850).

4 Davies, *Woman's Place Is at the Typewriter*, 18.

5 Strom, *Beyond the Typewriter*, 20.

6 Documented in ibid., 16.

7 Fine, *The Souls of the Skyscrapers*, 12.

8 Ibid.

9 Strom, *Beyond the Typewriter*, 26.

10 Davies, *Woman's Place Is at the Typewriter*, 71.

11 Fine, *The Souls of the Skyscraper*, 52.

12 Hilary Hallett, *Go West, Young Woman! The Rise of Hollywood* (Berkeley: University of California Press, 2013), 11.

13 Hilda Martindale, *Women Servants of the State, 1830–1938* (Oxford: Oxford University Press, 1988), 18.

14 Quoted in Strom, *Beyond the Typewriter*, 188.

15 Quoted in Fine, *The Souls of the Skyscrapers*, 22.

16 Jennifer L. Fleissner, "Dictation Anxiety: The Stenographer's Stake in Dracula," in *Literary Secretaries/Secretarial Culture*, 68.

17 Strom, *Beyond the Typewriter*, 175.

18 Fleissner, "Dictation Anxiety," 75.

19 Mary A. Laselle and Katherine E. Wiley, *Vocations for Girls* (New York: Houghton Mifflin, 1913), 103.

20 Alba M. Edwards, *Comparative Occupation Statistics for the United States, 1870–1940*, Part of the Sixteenth Census of the United States: 1940 (Washington, DC: Government Printing Office, 1943), tables 9 and 10, reprinted in Davies, *Women's Place Is at the Typewriter*, 178–179.

21 Kathleen Anne McHugh, *American Domesticity: From How-to Manual to Hollywood Melodrama* (Oxford: Oxford University Press, 1999), 76.

22 Quoted in ibid., 76.

23 Kessler-Harris, *Out to Work*, 233–234.

24 Fine, *The Souls of the Skyscrapers*, 60.

25 "Women's Sphere in the Business World," *Phonographic World*, March 1902, 14–23.

26 Quoted in Fleissner, "Dictation Anxiety," 83.

27 Lynn Peril, *Swimming in the Steno Pool* (New York: W. W. Norton & Company, 2011), 4.

28 Davies, *Woman's Place Is at the Typewriter*, 161–162.

29 M. Mostyn Bird, *Women at Work: A Study of the Different Ways of Earning a Living Open to Women* (London: Chapman and Hall, 1911), 135–136.

30 Kessler-Harris, *Out to Work*, 234.

31 Elizabeth Hillard Ragan, "One Secretary as per Specifications," *Saturday Evening Post*, December 12, 1931, 10.

32 Quoted in Kessler-Harris, *Out to Work*, 233–234.

33 Quoted in ibid., 234.

34 *Fortune*, August 1935, quoted in Fleissner, "Dictation Anxiety," 69.

35 E.g., Tom Gallon, *The Girl behind the Keys* (1903; reprint, Arlene Young, ed., Toronto: Broadview, 2006); Allen Grant, *Typewriter Girl* (1897; Broadview Encore Edition, Toronto: Broadview Press, Ltd., 2003).

36 Karen Ward Mahar, *Women Filmmakers in Early Hollywood* (Baltimore: Johns Hopkins University Press, 2006), 9–11, 14–15.

37 Charles Musser, *Before the Nickelodeon: Edwin S. Porter and the Edison Manufacturing Company* (Berkeley: University of California Press, 1991), 39.

38 Ibid., 51.

39 Staiger, *The Classical Hollywood Cinema*, 116.

40 Emile Colston, "Female Employment in Photography," *Photographic News* 32, no. 1547 (April 27, 1888): 266. Cited in discussion of female photo workers in Mahar, *Women Filmmakers*, 19–21. Further discussion of photographic work at studios later in this chapter.

41 Musser, *Before the Nickelodeon*, 41.

42 "The Art of Moving Picture Photography," *Scientific American*, April 17, 1897, 248–250. Practice was described in trade journals, as in "Selig's—The Great Moving Picture Plant of the West," *Moving Picture World* (hereafter referred to as *MPW*) 5, no. 8 (August 21, 1909): 247.

43 "Borrowed" in Joan Simon, "The Great Adventure," in *Alice Guy Blaché: Cinema Pioneer*, ed. Joan Simon (New Haven: Yale University Press, 2009), 5; "Arguably" in Anthony Slide, *The Silent Feminists: America's First Women Directors* (London: Scarecrow, 1996), vii.

44 Antonia Lant, "Introduction to Part Five," in *The Red Velvet Seat: Women's Writing on the First Fifty Years of Cinema* (London: Verso, 2006), 548–549.

45 Staiger, *The Classical Hollywood Cinema*, 117.

46 Ibid., 119.

47 Ibid., 116–117.

48 "Manager Hyman Increases Staff," *MPW* 22, no. 9 (November 28, 1914): 1220.

49 Charles Higham, *Cecil B. DeMille: A Biography of the Most Successful Film Maker of Them All* (New York: Charles Scribner's Sons, 1973), 39.

50 "Putting the Move in the Movies" recalls early production companies' disorganized purchasing agents and accounting practices, removed as they were from physical production. *Saturday Evening Post*, May 13, 1916, 14–15, 96–98, 100–101.

51 Discussed in Mahar, *Women Filmmakers*, 38.

52 Hallett, *Go West, Young Woman!*, 15.

53 Ibid., 13.

54 Adela Rodgers St. Johns, "Get Me Dorothy Arzner!" *Silver Screen* 4, no. 2 (December 1933): 24; see also Karyn Kay and Gerald Peary, "Interview with Dorothy Arzner," in *Women and the Cinema: A Critical Anthology*, ed. Kay and Peary (New York: E. P. Dutton, 1977), 154–155.

55 Cari Beauchamp, *Without Lying Down: Frances Marion and the Powerful Women of Early Hollywood* (Berkeley: University of California Press, 1997), 37.

56 Frances Marion, *Off with Their Heads: A Serio-Comic Tale of Hollywood* (New York: Macmillan, 1972), 13.

57 Beauchamp, *Without Lying Down*, 39.

58 DeWitt Bodeen, *More from Hollywood* (New York: A. S. Barnes and Company, 1977), 95.

59 Quoted in Kevin Brownlow, *The Parade's Gone By . . .* (Berkeley and Los Angeles: University of California Press, 1977), 276.

60 Anthony Slide, *Early Women Directors* (New York: A. S. Barnes and Company, 1977), 10.

61 Beauchamp, *Without Lying Down*, 11–12.

62 Janet Staiger summarizes these difficulties as described by a director to the *Saturday Evening Post* ("Putting the move in the movies," May 15, 1916, 14–15, 96–98, 100–101). *The Classical Hollywood Cinema*, 135.

63 Janet Wasko, *Movies and Money: Financing the American Film Industry* (New York: Ablex, 1982), 11–12.

64 For one discussion of professionalization that included all these emerging efficiency practices, see Louis Reeves Harrison, "Why Some Features Fail," *MPW* 22, no. 4 (October 24, 1914): 464.

65 Ibid. Systematization also discussed in "American Film Company Banquet," *MPW* 11, no. 2 (January 13, 1912): 121; "Recruits for Kinemacolor Company," *MPW* 15, no. 5 (February 1, 1913): 476.

66 "Studio Efficiency. Scientific Management as Applied to the Lubin Western Branch by Wilber Melville," *MPW* 17, no. 6 (August 9, 1913): 624.

67 "The Question of Efficiency in the Studio," *Motography* 14, no. 9 (August 28, 1915): 413. For rebuttal, see "The Efficiency Expert," *Motography* 14, no. 13 (September 25, 1915): 610; F. M. Taylor, "The Efficiency Man," *Motography* 14, no. 20 (November 13, 1915): 1010.

68 Taylor, "The Efficiency Man," 1010.

69 Davis, "Studio Efficiency," *MPW* 17, no. 6 (August 9, 1913): 624.

70 "H. O. Davis Talks System," *MPW* 28, no. 7 (May 13, 1916): 1142.

71 H. O. Stetchhan, "Efficiency in Studio Management," *Motography* 14.8 (August 21, 1915): 353.

72 Janet Staiger, "Dividing Labor for Production Control: Thomas Ince and the Rise of the Studio System," *Cinema Journal* 18, no. 2 (1979): 20.

73 Davis, "Studio Efficiency," 624.

74 "H. O. Davis Talks System," 1142.

75 Staiger, *The Classical Hollywood Cinema*, 134.

76 "H. O. Davis Talks System," 1142.

77 E. D. Horkheimer, "Studio Management," *MPW* 18, no. 6 (October 30, 1915): 982.

78 For efficiency backlash, see "Changes in Universal's Press Department," *MPW* 35, no.12 (March 23, 1918): 1689; Captain Leslie T. Peacocke, "Studio Conditions as I know them: Wanton Waste and Ignorant Efficiency, the Besettling Evils of Studio Management," *Photoplay,* vol. 12.1, (June 1917), 127.

79 For examples, see "Changes at Edison Studio," *Motography* 14.18 (October 30, 1915): 892; "Changes at Edison Studio," *MPW* 26, no. 5 (October 30, 1915): 768.

80 Here I borrow from John Caldwell who, in *Production Cultures*, unpacks DVD featurettes through a taxonomy that categorizes the bonus materials not by subject matter, but by the overall strategy or function they serve for the network, studio, or media makers who produced them. John Caldwell, *Production Culture: Industrial Reflexivity and Critical Practice in Film and Television* (Durham, NC: Duke University Press, 2008), appendix 2.

81 Early studios in houses discussed in "Kinematography in the United States," *MPW* 21, no. 2 (July 11, 1914): 177–179.

82 Eugene Dengler, "The Wonders of the 'Diamond-S' Plant," *Motography* 6, no. 1 (July 1911): 7–8.

83 For examples, see "Lubin Coronado Plant Opened," *MPW* 26, no. 2 (October 9, 1915): 236–237. George Blaisdell, "Where the Laughs Are Made," *MPW* 25, no. 2 (July 10, 1915): 233–234; "In the Field with Hotalling," *MPW* 15, no. 2 (January 9, 1913): 139; "Lubin Company Located," *MPW* 15, no. 6 (February 8, 1913): 560.

84 "Building Universal City," *MPW* 22, no. 1 (October 31, 1914): 49–50.

85 Ibid., 50.

86 Ince assets listed in: "Tom Ince and Inceville," *MPW* 21, no. 2 (July 11, 1914): 182; "Mecca of the Motion Picture," *MPW* 25, no. 2 (July 10, 1915): 216. "Forty-Three Acres for Incity," *MPW* 27, no. 6 (February 12, 1916): 958. "Ince to Move to Culver City," *MPW* 26, no. 2 (October 9, 1915): 272.

87 "Los Angeles Letter," *MPW* 25, no. 8 (August 21, 1915): 1301.

88 "From Forrest to Film," *Photoplay* 13, no. 5 (April 1918): 91.

89 Eugene Dengler, "The Wonders of the 'Diamond-S' Plant," *Motography* 6, no. 1 (July, 1911): 7–19.

90 "Scenes in Big Addition to Selig Studio," *Motography* 14, no. 14 (October 2, 1915): 674; "A partial view" in "Some Views of Metro's Gigantic New Studio," *Motography* 14, no. 11 (September 11, 1915): 510; see also "Scenes in the New Lubin Studio at Coronado," *Motography* 14, no. 15 (October 9, 1915): 728.

91 For examples of panoramas, see "Many Novelties in Studio Construction: Horseley Plant Unique," *Motography* 14, no. 7 (August 14, 1915): 291; "Balboa Studios Are Still Growing," *Motography* 14, no. 18 (October 30, 1915): 898. See also "Bird's-Eye View of Keystone Plant," *MPW* 25, no. 2 (July 10, 1915): 235.

92 "Motion Picture Colony under One Roof," *Scientific American* (June 21, 1919): 651.

93 Ibid.

94 "A Bird's Eye View of the Lasky Studio at Hollywood, California," *Photoplay* 13, no. 6 (May 1918): 30–31.

95 "Balboa Studios Are Still Growing," 898.

96 "Here's Efficiency for You," *Motography* 19, no. 15 (April 13, 1918): 707.

97 Mahar, *Women Filmmakers*, 194–195.

98 Staiger, *The Classical Hollywood Cinema*, 124, 134.

99 "Milton E. Hoffman, Studio Manager," *MPW* 35, no. 11 (March 16, 1918): 1487.

100 Davis, "Studio Efficiency," 624; Horkheimer, "Studio Management," 982; H. O. Stechhan, "Efficiency in Studio Management," *Motography* 14, no. 8 (August 21, 1915): 353. "H. O. Davis Talks System," 1142.

101 "Milton E. Hoffman, Studio Manager," *MPW* 35, no. 11 (March 16, 1918): 1487.

102 "Changes at Edison Studio," 892; see also "Centaur Activities," *MPW* 22, no. 2 (October 10, 1914): 175.

103 Photo spread in "The Executive Offices of V.L.S.E. Inc.," *Motography* 14, no. 3 (July 17, 1915): 104. New offices discussed in "V.L.S.E. Offices Enlarged," *Motography* 14, no. 9 (August 28, 1915): 395.

104 "Freuler Appoints Efficiency Men," *Motography* 14, no. 10 (September 4, 1915): 445.

105 "Metro Moving Offices," *Motography* 14, no. 26 (December 25, 1915): 1325.

106 Mark Garrett Cooper, *Universal Women: Filmmaking and Institutional Change in Early Hollywood* (Chicago: University of Illinois Press, 2010), 56–57, 63.

107 Staiger, *The Classical Hollywood Cinema*, 135.

108 "Notes of the Trade," *MPW* 18, no. 6 (November 8, 1913): 621.

109 Lubin in "Lubin Coronado Plant Opened," 236–237; Fox in "Motion Picture Colony Under One Roof," *Scientific American,* June 21, 1919, 651.

110 "A Bird's-Eye View of the Lasky Studio at Hollywood, California," 30–31.

111 Davis, "Studio Efficiency," 624.

112 M. Quadrelli, "Map of Universal City," circa 1914, Margaret Herrick Library, Academy of Motion Picture Arts and Sciences.

113 John Galtung, "A Structural Theory of Imperialism," *Journal of Peace Research* 8, no. 2 (1971): 81–117.

114 Quoted, discussed in Cooper, *Universal Women*, 58.

115 Horkheimer, "Studio Management," 982. Similar indexation described in "H. O. Davis Talks System," 1142.

116 Melvin M. Riddle, "From Pen to Silversheet V—Properties," *Photodramatist* 3, no. 11 (April 1922): 25–26.

117 Melvin M. Riddle, "From Pen to Silversheet III: Costuming the Players," *Photodramatist* 3, no. 9 (February 1922): 30.

118 Horkheimer, "Studio Management," 982; see also Melvin M. Riddle, "From Pen to Silversheet IX: Hunting Locations," *Photodramatist* 4, no. 3 (August 1922): 9.

119 "Changes at Lubinville: Hereafter Script Department to Take Its Proper Place in the Scheme of Things," *MPW* 16, no. 6 (May 24, 1913): 790.

120 "Casting Efficiency," *MPW* 26, no. 11 (December 11, 1915): 1985. Also in "Premier Simplifies Casting," *Motography* 14, no. 25 (December 11, 1915): 1284.

121 Melvin M. Riddle, "From Pen to Silversheet VI—Casting the Characters," *Photodramatist* 3, no. 12 (May 1922): 25–26.

122 As demonstrated by photograph of Mary Pickford dwarfed by towers of fan mail with the caption "Mary mailing a few pictures of herself to her admirers" in "Your Name, Please?" *Photoplay* 12, no. 4 (September 1917): 107.

123 "Levy Premier's Publicity Man," *MPW* 26, no. 11 (December 11, 1915): 1985.

124 "New Department Installed," *Motography* 18, no. 2 (July 14, 1917): 98.

125 R.V.S., "Scenario Construction," *MPW* 8, no. 6 (February 11, 1911): 294.

126 Mabel Condon, "What Happens to the Scenario," *Motography* 9, no. 2 (March 1, 1913): 147–151.

127 Ibid.

128 A Scenario Editor, "The Movies: A Colossus That Totters," *Bookman* 48, no. 6 (February 1919): 655.

129 "Changes at Lubinville," *MPW* 16, no. 6 (May 24, 1913): 790.

130 Ibid.

131 Ibid.

132 Jack G. Leo, "Greater Scenario Department," *MPW* 33, no. 3 (July 21, 1917): 382.

133 Captain Leslie T. Peacocke, "Logical Continuity," *Photoplay* 11, no. 5 (April 1917): 111.

134 Jeanie MacPherson, "Functions of the Continuity Writer," *Opportunities in the Motion Picture Industry* 3 (Los Angeles, Photoplay Research Society, 1922), 29.

135 H. O. Davis, "A Kitchener among Cameras," *Photoplay* 11, no. 6 (May 1917): 129–131, 147, 168–169.

136 Captain Leslie T. Peacocke, "Enter—the Free Lance Writer," *Photoplay* 11, no. 4 (March 1917): 97.

137 Davis, "A Kitchener among Cameras," 130.

138 Bradley King, "Some Studio Secrets," *Photodramatist* 3, no. 1 (June 1921): 19.

139 Kate Corbaley, "Duties and Qualifications of the Scenario Reader," *Opportunities in the Motion Picture Business* 2 (Los Angeles: Photoplay Research Society, 1922), 61–62.

140 Tino Balio, *Grand Design: Hollywood as a Modern Business Enterprise* (Berkeley: University of California Press, 1993), 99.

141 Beth Day, *This Was Hollywood: An Affectionate History of Hollywood's Golden Years* (New York: Doubleday, 1960), 228.

142 Ronald L. Davis, *The Glamour Factory: Inside Hollywood's Big Studio System* (Dallas, TX: Southern Methodist University Press, 1993), 162.

143 Balio, *Grand Design*, 99.

144 Selig discussed in Epes Winthrop Sargent, "The Photoplaywright," *MPW* 14, no. 1 (October 5, 1912): 38; Balboa in "The Balboa Enterprise," *MPW* 25, no. 2 (July 10, 1915): 246; Universal in Davis, "A Kitchener among Cameras," 147.

145 Lasky discussed in Melvin M. Riddle, "From Pen to Silversheet II—Architecture, Decoration, Research," *Photodramatist* 3, no. 8 (January 1922): 37; MGM in Steven Bingen, Stephen X. Sylvester, and Michael Troyen, *M-G-M: Hollywood's Greatest Backlot* (Solana Beach, CA: Santa Monica Press, 2011), 59.

146 Balio, *Grand Design*, 87.

147 Riddle, "From Pen to Silversheet II," 37.

148 Bingen et al., *M-G-M*, 59.

149 Davis, "A Kitchener among Cameras," 130.

150 Howard Koch, *As Time Goes By: Memoirs of a Writer* (New York: Harcourt, 1979), 43.

151 Briefs described in Melvin M. Riddle, "From Pen to Silversheet V—Properties," *Photodramatist* 3, no. 11 (April 1922): 25–26. Full process at Warners/"bibles" discussed in Carl Milliken, "Information Please," *Warner Club News* (June 1940): 3.

152 Legal discussed in Samuel Marx, *A Gaudy Spree: Literary Hollywood When the West Was Fun* (New York: Franklin Watts, 1987), 119; Day, *This Was Hollywood*, 237.

153 PCA liaisons and scripting in Ruth Vasey, *The World according to Hollywood, 1918–1939* (Madison: University of Wisconsin Press, 1997), 131–132, 135–136, 164–166.

154 Anita Loos and John Emerson, "How to Write Movies," *Photoplay* 17, no. 2 (February 1920): 50; Marion stenographer in Beauchamp, *Without Lying Down*, 79.

155 "Los Angeles Letter," *MPW* 25, no. 8 (August 21, 1915): 1301.

156 Melvin M. Riddle, "From Pen to Silversheet," *Photodramatist* 3, no. 7 (December 1921): 15–17.

157 Ibid., 17.

158 For a description of these departments at MGM, see Marx, *A Gaudy Spree*, 9–10, 91.

159 Since 1941, the Writer's Guild of America has been the final arbiter of screen credit. Their designations are described in their Screen Credits Manual, http://www.wga.org/subpage_writersresources.aspx?id=167. Explanation of basic credit designations can be found at http://www.imdb.com/partners/wga.

Chapter 2 Studio Tours:
Feminized Labor in the Studio System

1 Karen Ward Mahar, "Doing a Man's Work," in *The Classical Hollywood Reader*, ed. Steven Neale (London: Routledge, 2012), 83.

2 Janet Staiger, *The Classical Hollywood Cinema*, by David Bordwell, Janet Staiger, and Kristin Thompson (New York: Columbia University Press, 1985), 106, 212.

3 Karen Ward Mahar, *Women Filmmakers in Early Hollywood* (Baltimore: Johns Hopkins University Press, 2006), 180.

4 Karen Kay and Gerald Peary, eds., *Women and the Cinema: A Critical Anthology* (New York: E. P. Dutton, 1977), 158.

5 Quoted in Cari Beauchamp, *Without Lying Down: Frances Marion and the Powerful Women of Early Hollywood* (Berkeley: University of California Press, 1997), 352.

6 Ibid., 355.

7 Frances Marion, interviewed in DeWitt Bodeen, *More from Hollywood!: The Careers of 15 Great American Stars* (New York: A. S. Barnes and Company, 1977), 113.

8 Lizzie Francke, *Script Girls: Women Screenwriters in Hollywood* (London: BFI, 1994), 41.

9 Frances Marion, *Off with Their Heads: A Serio-Comic Tale of Hollywood* (New York: Macmillan, 1972), 277–278.

10 Bodeen, *More from Hollywood*, 116.

11 Kay and Peary, *Women and the Cinema*, 152.

12 Beauchamp, *Without Lying Down*, 41.

13 Myrtle Gebhart, "Her Film Hits Open the Way for More Women Producers," *Boston Sunday Post*, August 13, 1944. Quoted in Francke, *Script Girls*, 60.

14 Martha M. Lauzen, "The Celluloid Ceiling: Behind-the-Scenes Employment of Women on the Top 250 Films of 2013." Study Conducted by the Center for the Study of Women in Television and Film (San Diego, CA: San Diego State University, 2014), http://womenintvfilm.sdsu.edu/research.html.

15 Organisation for Economic Co-Operation and Development, *The Future of Female-Dominated Occupations* (Paris: Author, 1998), 9.

16 *Behind the Screen*, excerpt, directed by Al Christie, Universal, 1915, UCLA Film and Television Archive, catalog #VA11747 M. *Universal Studio and Stars*, Universal, 1925, UCLA Film and Television Archive, catalog #VA19365 M.

17 *A Tour of the Thomas H. Ince Studio, 1920–22*, directed by Hunt Stromberg, Ince Studio, 1924, accessed at UCLA Film and Television Archive, catalog #VA2945 M; *A Trip to Paramountown*, directed by Jack Cunningham, Paramount, 1922, UCLA Film and Television Archive, catalog # VA4707 M.

18 *1925 Studio Tour*, MGM, 1925, UCLA Film and Television Archive, catalog #VA13208 M.

19 *Life in Hollywood*, directed by L. M. Be Dell, Goodwill Pictures, Inc., 1927. http://youtu.be/EPLMHHZeKqE. The films *Warner Bros Studios and Stars* (William Horsley, producer, made between 1923 an 1927, UCLA Film and Television Archive, catalog #VA22187) and *William Fox Studio and Stars*

(William Horsley, producer, made between 1915 and 1927, UCLA Film and Television Archive, catalog #VA19592 M) each contain footage from this film.

20 *A Trip thru a Hollywood Studio*, directed by Ralph Staub, Vitaphone, 1934, UCLA Film and Television Archive, catalog # M108519; *20th Century Fox Tour with Darryl F. Zanuck*, 20th Century Fox, 1935, UCLA Film and Television Archive, catalog #VA4480 M.

21 Anthony Slide, *Early Women Directors* (New York: A. S. Barnes and Company, 1977), 9.

22 A completely accurate count is not possible due to picture and cropping of the frame.

23 Female screenwriter Bradley King was employed at Ince in the early 1920s when the Ince studio tour was supposedly shot. See Bradley King, "More Studio Secrets," *Photodramatist* 3, no. 12 (May 1922): 5–6.

24 Jane Gaines and Radha Vatsal, "How Women Worked in the US Silent Film Industry," in *Women Film Pioneers Project, ed.* Jane Gaines, Radha Vatsal, and Monica Dall'Asta, Center for Digital Research and Scholarship. New York: Columbia University Libraries, 2013. Web. November 18, 2011. https://wfpp.cdrs .columbia.edu/essay/how-women-worked-in-the-us-silent-film-industry/.

25 Mark Garrett Cooper, *Universal Women: Filmmaking and Institutional Change in Early Hollywood* (Chicago: University of Illinois Press, 2010), 44.

26 Danae Clark, *Negotiating Hollywood: The Cultural Politics of Actors' Labor* (Minneapolis: University of Minnesota Press, 1995), 19.

27 Mahar, *Women Filmmakers*, 196–197.

28 Quoted in Howard Koch, *As Time Goes By: Memoirs of a Writer* (New York: Harcourt, 1979), 24.

29 John Caldwell examines trade narratives of various crafts in *Production Culture: Industrial Reflexivity and Critical Practice in Film and Television* (Durham, NC: Duke University Press, 2008), while Felicia Henderson discusses the casting of television writing staffs in Henderson, "Both Sides of the Fence: The Writer's Room," in *Production Studies: Cultural Studies of Media Industries*, ed. Vicki Mayer, Miranda J. Banks, and John Thornton Caldwell (London: Routledge, 2009), 224–229.

30 Described by David Stenn in "It Happened One Night . . . At MGM," *Vanity Fair*, April 2003, http://www.vanityfair.com/fame/features/2003/04/mgm200304.

31 Alice Kessler-Harris, *Out to Work: A History of Wage-Earning Women in the United States* (1982; rpt., 20th anniv. ed., Oxford: Oxford University Press, 2003), 128–137.

32 Ibid.

33 Mary A. Laselle and Katherine E. Wiley, *Vocations for Girls* (New York: Houghton Mifflin, 1913), 103.

34 Beth Day, *This Was Hollywood: An Affectionate History of Filmland's Golden Years* (New York: Doubleday and Company, 1960), 19.

35 Christopher Finch and Linda Rosenkrantz, *Gone Hollywood: The Movie Colony in the Golden Age* (New York: Doubleday, 1979), 68.

36 Russell interviewed in Mike Steen, *Hollywood Speaks: An Oral History* (New York: G. P. Putnam and Sons, 1974), 100.

37 Steven Bingen, Stephen X. Sylvester, and Michael Troyen, *M-G-M: Hollywood's Greatest Backlot* (Solana Beach, CA: Santa Monica Press, 2011), 99.

38 Custodians discussed in "Personal from Personnel," *MGM Studio Club News* (May 8, 1940): 23. For examples of custodian columns, see Louise Johnson, "Custodians," *Warner Club News* (May 1944): 13.

39 Horace Hampton, "Sweepings," *Warner Club News* (January 1940).

40 "Custodians," *Warner Club News* (1941); Custodians and Service Dance photograph in *Warner Club News* (October 1941).

41 "20th Century-Fox Workflow Diagrams—Service Department," *Infomercantile.com*, circa 1940s, http://www.infomercantile.com/-/ 1940s_Studio_Organization,_20th_Century_Fox.

42 Ted Crumley, "Sweepings," *Fox Close-Ups* (April 1937): 2.

43 "Metro-Goldwyn-Mayer," *Fortune*, December 1932, 51–58; reprinted in Tino Balio, ed., *The American Film Industry* (Madison: University of Wisconsin Press, 1985), 311.

44 Finch and Rosenkrantz, *Gone Hollywood*, 88.

45 Bingen, Sylvester, and Troyen, *M-G-M*, 64.

46 Barbershops existed (and still exist) at Paramount and Universal, and are described at Warner Bros. in E. J. Stephens and Marc Wanamaker, *Early Warner Bros. Studios* (Charleston, SC: Arcadia, 2010), 51.

47 Image appears in *Warner Club News* (October 1940): 9.

48 Beauty industry as masculinized in Mahar, *Women Filmmakers*, 203; Westmore dominance discussed in Finch and Rosenkrantz, *Gone Hollywood*, 183–185.

49 Mervyn LeRoy, *It Takes More Than Talent* (New York: Alfred A. Knopf, 1953).

50 Day, *This Was Hollywood*, 171; Linda Seger, *When Women Call the Shots: The Developing Power and Influence of Women in Television and Film* (New York: Henry Holt, 1996), 14.

51 Ronald L. Davis, *The Glamour Factory: Inside Hollywood's Big Studio System* (Dallas, TX: Southern Methodist University Press, 1993), 223–225.

52 Steen, *Hollywood Speaks*, 276, 284.

53 Fay Wray, "Women in Film Oral History Interview of Fay Wray by Andrea S. Walsh," Women in Film Foundation, UCLA Film and Television Archive, July 24–28, 1989, 114.

54 Julie Lugo Cerra and Marc Wanamaker, *Movie Studios of Culver City* (Charleston, SC: Arcadia, 2011), 26.

55 Finch and Rosenkrantz, *Gone Hollywood*, 61. Zanuck friendship discussed in Davis, *The Glamour Factory*, 318.

56 Davis, *The Glamour Factory*, 319.

57 Finch and Rosenkrantz, *Gone Hollywood*, 57–58.

58 Warners commissary pictures in Stephens and Wanamaker, *Early Warner Bros. Studios*, 51, 94.

59 Julie Lang Hunt, "They Aren't All Actresses in Hollywood," *Photoplay* (September 1936): 92.

60 "20th Century-Fox Workflow Diagrams—Restaurants," *Infomercantile.com*, circa 1940s, http://www.infomercantile.com/-/1940s_Studio_Organization, _20th_Century_Fox.

61 "Commissary Unit Forms Only Girls' Bowling Team and Defeats Males!" *MGM Studio Club News* (May 5, 1938): back cover.

62 "Café Built on Lot 2," *MGM Studio Club News* (February 8, 1938): front cover.

63 For example, see Cerra and Wanamaker, *Movie Studios of Culver City*, 43.

64 Bingen et al., *M-G-M*, 44.

65 Restaurant and hotel dining hall management discussed as viable fields in Miriam Simons Leuck, *Fields of Work for Women* (New York: D. Appleton and Company, 1926), 74–75. Personnel management as "most 'feminine'" of business professions in Sharon Hartman Strom, *Beyond the Typewriter: Gender, Class, and the Origins of Modern American Office Work, 1900–1930* (Chicago: University of Illinois Press, 1992), 81.

66 "20th Century-Fox Workflow Diagrams—Restaurants."

67 Finch and Rosenkrantz, *Gone Hollywood*, 60.

68 Davis, *The Glamour Factory*, 318–319.

69 "Boss of the commissary" in Morris Gardner, "Commissary Gossip," *MGM Studio Club News* (November 18, 1937): 4 (pages not numbered). "commissary cuties" in Norman Farer, "Commissary Quickies," *MGM Studio Club News* (May 1940): 18.

70 "20th Century-Fox Workflow Diagrams—Restaurants,"

71 Fred Wagner, "What It Takes," *Warner Club News* (April 1937): 1.

72 Davis, *The Glamour Factory*, 316.

73 Ibid., 313.

74 The field was characterized as a women's field by Edith Clark of the Christie Film Company in 1922, but there is little evidence that women supervised costume departments. Edith Clark, "Designing Clothes for Movie Folk," *Opportunities in the Motion Picture Industry*, no. 3 (Los Angele: Photoplay Research Society, 1922), 79–81.

75 Male domination of art direction discussed in Slide, *Early Women Directors*, 10–11 and Mahar, *Women Filmmakers*, 196.

76 Seger, *When Women Call the Shots*, 10.

77 Finch and Rosenkranz, *Gone Hollywood*, 68–72. For further evidence, see costuming chapters in Davis, *The Glamour Factory*; Day, *This Was Hollywood*; or Tino Balio, *Grand Design: Hollywood as a Modern Business Enterprise, 1930–39* (Berkeley: University of California Press, 1993).

78 Balio, *Grand Design*, 92.

79 Wagner, "What It Takes," 1.

80 Melvin M. Riddle, "From Pen to Silversheet IV—Filmland's Fashion Shop," *Photodramatist* 3, no. 10 (March 1922): 31.

81 Finch and Rosenkrantz, *Gone Hollywood*, 71.

82 One such photo of MGM wardrobe department can be found at digital gallery ID # 1084_049613, Core Collection, Subject Files, Margaret Herrick Library, Los Angeles, CA.

83 For a description of this feminization process, see Harris, *Out to Work*, 21–44.

84 For example, female workers also pictured in this posture in the wardrobe department at Lubin Coronado in "Scenes in the New Lubin Coronado Studio at Coronado," *Motography* 14, no. 15 (October 9, 1915): 728.

85 Strauss quoted in Davis, *The Glamour Factory*, 216.

86 Day, *This Was Hollywood*, 128–129.

87 Photographic manufacture described in Mahar, *Women Filmmakers*, 20–21. Early motion picture work in Charles Musser, *Before the Nickelodeon: Edwin S. Porter and the Edison Manufacturing Company* (Berkeley and Los Angeles: University of California Press, 1991), 41.

88 "Selig's The Great Moving Picture Plant of the West," *Moving Picture World* 5, no. 8 (August 21, 1909): 248. Eugene Dengler, "The Wonders of the Diamond-S Plant," *Motography* 6, no. 1 (July 1911). "Scenes in Big Addition to Selig Studios," *Motography* 14, no. 14 (October 2, 1915): 674.

89 Melvin Riddle, "From Pen to Silversheet XIV—The Film Laboratory," *Photodramatist* 4, no. 9 (February 1923): 6, 38.

90 Mahar, *Women Filmmakers*, 24.

91 Dmytryk, *It's a Hell of a Life, but Not a Bad Living* (New York: Times Books, 1978), 17–18.

92 "1938 Rejection Letter from Disney to a Female Artist," http://holykaw.alltop .com/1938-rejection-letter-from-disney-to-a-female-a-female-artist.

93 Patricia Zohn, *Coloring the Kingdom, Vanity Fair,* March 2010, http://www .vanityfair.com/culture/features/2010/03/disney-animation-girls-201003.

94 Ibid.

95 Tom Sito, *Drawing the Line: The Untold Story of Animation Unions from Bosko to Bart Simpson* (Lexington: University of Kentucky Press, 2006), 26.

96 "Cartoons," *MGM Studio Club News* (November 18, 1937): 2.

97 Martha Sigall, *Living Life inside the Lines: Tales from the Golden Age of Animation* (Jackson: University Press of Mississippi, 2005), 110–115. There are also frequent mentions of "beauties" in the cartoon department columns of the *Warner Club News*. For examples, see July 1936 and April 1942 issues.

98 Sigall, *Living Life inside the Lines*, 110.

99 Zohn, *Coloring the Kingdom*.

100 LeRoy, *It Takes More Than Talent*, 1953.

101 Sito, *Drawing the Line*, 336.

102 Harris, *Out to Work*, 127.

103 Laselle and Wiley, *Vocations for Girls*, 73.

104 Dr. Helen Jones, "Notes from the Doctor's Chart," *MGM Studio Club News* (May 8, 1940): 9.

105 Bingen et al., *M-G-M*, 78. Day, *This Was Hollywood*, 176–177.

106 For description of department's nurses at RKO see "Immateria Medica," *RKO Studio News* (October 1935): 33.

107 See for example, Warner Bros. school pictured in 1939: Stephens and Wanamaker, *Early Warner Bros. Studios*, 77.

108 Bingen et al., *M-G-M*, 77.

109 Ida R. Koverman, "The Little Green Room," *MGM Studio News* (December 1941): 6–7.

110 Described in Davis, *The Glamour Factory*, 85–93.

111 Finch and Rosenkrantz, *Gone Hollywood*, 45–49.

112 Davis, *The Glamour Factory*, 132.

113 For example, men were more heavily concentrated in the areas of piano and singing, while women often taught voice, drama, and ballet.

114 Frank Schwartzman and Les Martinson, "Front Office Flashes," *MGM Studio Club News* (November 18, 1937): 4.

115 Bingen et al., *M-G-M*, 38.

116 Fred Pappmeier, "Studio Tours: Timekeeping," *Warner Club News* (February 1937), 1.

117 Fred Wagner, "What's What and Who's Who on the Backlot: Timekeeping," *Warner Club News* (February 1937): 7. Brain machine described in Fred Pappmeier, "Studio Tours," 1.

118 Wagner, "What's What and Who's Who on the Backlot: Tabulating Department," *Warner Club News*, (February, 1937): 10.

119 Robert L. Boggs, "Telegraph," *MGM Studio Club News* (October 1942): 8.

120 Ellen Lupton, *Mechanical Brides: Women and Machines from Home to the Office* (New York: Princeton Architectural Press, 1993).

121 Jean Booth, "PHONE in Lines," *MGM Studio Club News* (December 24, 1938): 8.

122 Berman interview in Steen, *Hollywood Speaks*, 168–169.

123 Davis, *The Glamour Factory*, 138.

Chapter 3   The Girl Friday and How She Grew: Female Clerical Workers and the System

1 "Studio Romance Disclosed: Ring Lardner, Jr., to Marry Secretary of Selznick," *Los Angeles Times*, February 5, 1937.

2 "They Aren't All Actresses in Hollywood," *Photoplay* 50, no. 3 (September 1936): 92.

3 Catherine Filene, ed., "The Motion-Picture Industry: Outline of Opportunities in the Motion Picture Occupations," in *Careers for Women* (Boston: Houghton Mifflin, 1934), 432–433.

4 Ida May Park, "Motion Picture Work: The Motion-Picture Director," in *Careers for Women*, ed. Filene, 335–337.

5 *MGM Studio Club News* (June 1936).

6 May Wale Brown, *Reel Life on Hollywood Movie Sets* (Los Angeles: Ariadne, 1995), 3.

7 H. O. Stetchhan, "Efficiency in Studio Management," *Motography* 14, no. 8 (August 21, 1915): 353.

8 Richard Koszarski, *An Evening's Entertainment: The Age of the Silent Feature Picture, 1915–1928* (Berkeley: University of California Press, 1990), 109.

9 Roger Mayer, quoted in Steven Bingen, Stephen X. Sylvester, and Michael Troyen, *M-G-M: Hollywood's Greatest Backlot* (Solana Beach, CA: Santa Monica Press, 2011), 22.

10 Quoted in Ronald L. Davis, *The Glamour Factory: Inside Hollywood's Big Studio System* (Dallas, TX: Southern Methodist University Press, 1993), 313.

11 Robert Parrish, *Growing Up in Hollywood* (New York: Harcourt Brace Jovanovich, 1976), 119.

12 Leo Rosten, *Hollywood: The Movie Colony, the Movie Makers* (New York: Harcourt Brace, 1941), 246.

13 "Studio Efficiency," *Moving Picture World* (hereafter *MPW*) 17, no. 6 (August 9, 1913): 624.

14 Thomas H. Ince, "Ince Makes War on Inconsistency: Producer, Citing Many Errors in Films Generally, Tells How He Keeps Them Out of Own Pictures," *Motography* 19, no. 8 (February 23, 1918): 361; "Devotion to Detail at Lasky's," *Photoplay* 14, no. 1 (June 1918): 34; E. H. Calvert, "Attention to Detail Makes for Success," *Motography* 14, no. 19 (November 6, 1915): 947.

15 H. O. Davis, "A Kitchener among Cameras," *Photoplay* 11, no. 6 (May 1917): 130.

16 King Vidor, *King Vidor on Film Making* (New York: David McKay Company, 1972).

17 Producer numbers in Rosten, *Hollywood*, 246.

18 Quoted in Tino Balio, *Grand Design: Hollywood as a Modern Business Enterprise, 1930–39* (Berkeley: University of California Press, 1993), 79.

19 Ibid., 84.

20 Beth Day, *This Was Hollywood: An Affectionate History of Filmland's Golden Years* (New York: Doubleday, 1960), 156, 134, 128–129.

21 Anthony Slide, *Silent Topics* (London: Scarecrow Press, 2005), 23.

22 Twenty percent figure in Slide, *Silent Topics*, 23–27; 1930s increase detailed in Koszarski, *An Evening's Entertainment*, 119.

23 Vidor, *King Vidor on Film Making*, 14.

24 Hortense Powdermaker, *Hollywood: The Dream Factory* (New York: Little Brown, 1950), 88–90.

25 Department names and organization drawn from various sources in which studio hierarchies are comprehensively described, including material from the Warner Personnel Files in the University of Southern California Warner Archive, Davis's *The Glamour Factory*, which describes the studios from the top down, an organizational chart from RKO in 1934 reprinted in Richard Jewell's *The Golden Age of Cinema: Hollywood, 1929–45* (London: Wiley Blackwell, 2007); "20th Century–Fox Workflow Diagrams," *Infomercantile.com*, circa 1940s, http://www.infomercantile.com/-/1940s_Studio_Organization,_20th_Century_Fox; and "What It Takes," Fred Wagner's 1937 *Warner Club News* article cataloguing of the jobs done by the Warner Bros. three thousand below-the-line employees (April 1937: 1).

26 Miriam Simons Leuck, *Fields of Work for Women* (New York: D. Appleton and Company, 1926), 56–57.

27 Ibid., 8.

28 Edward Dmytryk, *It's a Hell of a Life but Not a Bad Living: A Hollywood Memoir* (New York: Times Books, 1978), 3–4.

29 The election was ordered by the National Labor Relations Board (hereafter referred to as NLRB) following an investigation based on petitions filed by employees at the studios. "Columbia Pictures Corporation, Samuel Goldwyn, Inc., LTD., Loews's, Inc., R.K.O. Radio Pictures, Inc., Republic Productions, Inc., Hal Roach Studios, Inc., Selznick International Pictures, Inc., 20th Century Fox Film Corporation, Universal Pictures Company, Inc., Walter Wanger Productions, Ink, and Screen Office Employees Guild (Ind.)," Cases Nos. R-2035 to R-044 inclusive, National Labor Relations Board, Decided October 8, 1940; Election results

reported in "SOEG Victor in Studios' Balloting: NLRB Election Decides Problem," *RKO Studio News* (November 1940), 2.

30 20th Century-Fox Film Corporation and Screen Office Employees Guild, Local 1391, A.F.L., Case no. 21-R-2252, National Labor Relations Board, Decided April 29, 1944.

31 "SOEG Victor in Studios' Balloting: NLRB Election Decides Problem," *RKO Studio News* (November 1940): 2.

32 Fred Wagner, "What It Takes," *Warner Club News* 2, no. 2 (April 1937), 1.

33 The rating system and pay scale is explained in "Initial Rating Record," USC Warner Bros. Archive, Personnel Files Box 380b, undated, and "Rating Ledger," USC Warner Bros. Archive, Personnel Files Box 380b, undated.

34 I determined workers' gender based on their first names and any other available information. This is not an exact science but gives a roughly accurate count.

35 "Job Titles and Classifications," USC Warner Bros. Archive, Personnel Files Box 380b, dated November 1949.

36 Ibid.

37 Purchasing department pictured in Bingen et al., *M-G-M*, 32. For pictures of accounting department circa 1944 see digital ID# 1084_072090, Core Collection, Margaret Herrick Library, Los Angeles, CA.

38 "20th Century—Fox Workflow Diagrams—Legal Department," *Infomercantile.com*, circa 1940s, http://www.infomercantile.com/-/1940s_Studio_Organization,_20th_Century_Fox.

39 Interviewed in Karen Kay and Gerald Peary, eds., *Women and the Cinema* (New York: E. P. Dutton, 1977), 154–155.

40 "Interviewing Helen Gregg," *RKO Studio Club News* (February 1941): 8.

41 Ibid.

42 Described in "Interviewing Wynne Haslam," *RKO Studio Club News* (April 1941): 9, 12.

43 Ibid., 9.

44 "All This and Sadie, Too," *Warner Club News* (November 1940): 3, 12.

45 Ibid.

46 "Typewriter Department," *Warner Club News* (May 1940): 9.

47 Cari Beauchamp, ed., *Adventures of a Hollywood Secretary: Her Private Letters from inside the Studios of the 1920s* (Berkeley: University of California Press, 2006), 167.

48 Valeria Beletti, letter to Irma Prina, December 15, 1928, in Beauchamp, ed., *Adventures of a Hollywood Secretary*, 200.

49 Davis, *Glamour Factory*, 168.

50 Mervyn LeRoy, *It Takes More Than Talent* (New York: Alfred A. Knopf, 1953), 149.

51 Ibid.

52 *A Gaudy Spree: Literary Hollywood When the West Was Fun* (New York: Franklin Watts, 1987), 10.

53 Bingen et al., *M-G-M*, 22.

54 Wilde, *Loving Gentleman*, 86.

55 "20th Century–Fox Workflow Diagrams—Script Department," *Infomercantile.com*, circa 1940s, accessed online at: http://www.infomercantile.com/-/1940s_Studio_Organization,_20th_Century_Fox.

56 "All This and Sadie, Too," *Warner Club News* (November 1940): 3, 12.

57 "Interviewing Wynne Haslam," *RKO Studio Club News* (April 1941): 9.

58 "All This and Sadie, Too," 12.

59 "Secretaries," *RKO Studio Club News* (April 1941): 4.

60 Marion Snell, "Secretaries," *RKO Studio Club News* (December 1940): 15.

61 Ezra Goodman, "How to Be a Hollywood Producer," *Harper's* (May 1948): 415–423.

62 "Secretary to Producer—Job Analysis," USC Warner Bros. Archive, Personnel Files Box 380 B, April 8, 1947.

63 Davis, *Glamour Factory*, 312.

64 Day, *This Was Hollywood*, 236–237.

65 Ibid., 131–132.

66 The Billy Gordon Papers at the Margaret Herrick Library contain hundreds of examples of these casting lists, suggestions, and discussions as they were circulated in memo form. For a typical example, see the cast lists and casting suggestions for *Kiss of Death* (1947, 20th Century–Fox). William "Billy" Gordon Papers, KISS OF DEATH Casting—Folder 4.f-158, Margaret Herrick Library Special Collections, Los Angeles, CA.

67 Robert Vogel, *Robert Vogel Oral History*. Interview by Barbara Hall, Margaret Herrick Library, 1990, 77.

68 Davis, *Glamour Factory*, 143.

69 Day, *This Was Hollywood*, 196–197.

70 "Job Titles and Classifications," USC Warner Bros. Archive, Personnel Files, Box 380b, dated November 1949.

71 "Department Clerk, Makeup—Job Analysis," USC Warner Bros. Archive, Personnel Files, Box 380B, July, 1952. "Jr. Clerk, Special Photographic Effects & Matte—Job Description," USC Warner Bros. Archive, Personnel Files, Box 380B, undated.

72 Mike Steen, *Hollywood Speaks: An Oral History* (New York: G. P. Putnam and Sons, 1974).

73 "Promote from the Ranks," *MGM Studio Club News*, 3, no. 5 (May 9, 1938): 2.

74 Temp agency submissions: "Performance Personnel Resumes," USC Warner Bros. Archive, Personnel Files Box 380a.

75 For example, see: "Secretary, H. M. Warner Office—Job Analysis," USC Warner Bros. Archive, Personnel Files, Box 382 B, June 11, 1947. This range of requirements also discussed in LeRoy, *It Takes More Than Talent*, 153.

76 Leah Price and Pamela Thurschwell, "Introduction: Invisible Hands," in *Literary Secretaries/Secretarial Culture*, ed. Leah Price and Pamela Thurschwell (Hampshire, UK: Ashgate Publishing, 2005), 5.

77 Susan Elizabeth Dalton, "Women at Work: Warners in the 1930s," in *Women and the Cinema*, ed. Karen Kay and Gerald Peary (New York: E. P. Dutton, 1977), 276–277.

78 Lynn Peril, *Swimming in the Steno Pool: A Retro Guide to Making It in the Office* (New York: W. W. Norton, 2011), 44.

79 Schulman, *I Lost My Girlish Laughter*, 140.

80 For examples, see Rose Davidson, "Secretaries," *Warner Club News* (January 1946), 8, and Thelma Hanover, "Stenographic," *Warner Club News* (July 1946): 18.

81 "Reading Department," *MGM Studio Club News* (December 1941).

82 "Notes from Stenographic," *RKO Studio Club News* 2, no. 6 (November 1939): 2–3; poll results described in Marion Dix, "Secretaries," *Warner Club News* (January 1944): 4.

83 "Writer," *Warner Club News* (March 1937): 7.

84 Fred Wagner, "What's What and Who's Who on the Backlot," *Warner Club News* (February 1937): 10.

85 For secretaries as pinups, see *MGM Studio Club News* (June 1944); for Hollywood Canteen, see Ruth Ellen Moore, "Cartoons," *MGM Studio Club News* (October 1942): 14; Hedda Hopper, "Hedda Hopper's Hollywood," *Los Angeles Times* (November 17, 1942); "Studios to Entertain Soldiers at Party," *Los Angeles Times* (November 11, 1941); for pen pal program see *MGM Studio Club News* (June 1941): 20.

86 "Untitled," *Warner Club News* (February 1937): 7.

87 Charles Rycroft, *A Critical Dictionary of Psychoanalysis*, 2nd ed. (London: Penguin Reference Books, 1995).

88 "Hairdress" in Silvia Rosenthal, "Secretaries," *Warner Club News* (August 1939): 6; "WANTED" in Joan Dawson, "Secretaries," *Warner Club News* (February 1940): 10.

89 Lizzie Francke, *Script Girls: Women Screenwriters in Hollywood* (London: BFI, 1994), 59, 62.

90 Ibid., 59–60.

91 Quoted in Ellen Lupton, *Mechanical Brides: Women and Machines from Home to the Office* (New York: Princeton Architectural Press, 1993), 48.

92 For descriptions of such tactics (e.g., personnel kept waiting, psychology of office layouts of Goldwyn, Mayer, and Cohn), see Christopher Finch and Linda Rosenkrantz, *Gone Hollywood: The Movie Colony in the Golden Age* (New York: Doubleday, 1979), 224–225.

93 Bennett Quoted in Patrick McGilligan, *Backstory: Interviews with Screenwriters of Hollywood's Golden Age* (Berkeley: University of California Press, 1986), 36. Further discussion by Francke, *Script Girls*, 56.

94 Tom Sito, *Drawing the Line: The Untold Story of Animation Unions from Bosko to Bart Simpson* (Lexington: University of Kentucky Press, 2006), 20.

95 Valeria Beletti, Letter to Irma Prina, February 19, 1925, in Beauchamp, ed., *Adventures of a Hollywood Secretary*, 17.

96 Ronald Davis gives this figure in *The Glamour Factory*, 312. Torchia quoted in Davis, *The Glamour Factory*, 312.

97 For example of Torchia columns, see "Style Scoops," *MGM Studio Club News* 3 (August 12, 1938): 4; Renie, "Fashion's Fancy," *RKO Studio Club News* (February, 1941): 16.

98 Ibid., 60–61.

99 Marx, *A Gaudy Spree*, 28.

100 Ibid., 29.

101 Meta Carpenter Wilde and Orin Borsten, *A Loving Gentleman: The Love Story of William Faulkner and Meta Carpenter* (New York: Simon and Schuster, 1976), 94–95.

102 Ibid.

103 Here I borrow John Caldwell's concept of "para-industries," which describes subcontractors in the present age of post-Fordist production outsourcing. See "Para-Industries: Researching Hollywood's Blackwaters," *Cinema Journal* 52, no. 3 (Spring 2013): 157.

104 Ibid.

105 Brown, *Reel Life*, 177.

106 Richard Lemon, "Queens of Gossip," *People*, May 13, 1985, 133.

107 Eleanor Gilbert, *The Ambitious Woman in Business* (New York: Funk and Wagnall's, 1916), 141.

108 Peril, *Swimming in the Steno Pool*, 161–163.

109 Ibid., 164.

110 Valeria Beletti, Letter to Irma Prina, April 8, 1925, in Beauchamp, ed., *Adventures of a Hollywood Secretary*, 34.

111 Valeria Beletti, Letter to Irma Prina, May 20, 1925, in Beauchamp, ed., *Adventures of a Hollywood Secretary*, 45–46.

112 Marcella Rabwin, *Yes, Mr. Selznick: Recollections of Hollywood's Golden Era* (Pittsburg, PA: Dorrance Publishing, 1999), 153.

113 Ibid., 164.

114 Ibid,. 164.

115 Wilde and Borsten, *A Loving Gentleman*, 37.

116 Alma Young, interview by Anthony Slide and Robert Gitt, Margaret Herrick Library, 1977, 7.

117 Valeria Beletti, Letter to Irma Prina, June 10, 1925, in Beauchamp, ed., *Adventures of a Hollywood Secretary*, 52–53.

118 Marx, *A Gaudy Spree*, 85–86.

119 "Man to Face Party Girl: Showdown Near in Film Convention Hayloft Revel," *Los Angeles Times*, June 6, 1937.

120 *MGM 1937 Convention*, MGM 1937, UCLA Film and Television Archive, catalog #M38478.

121 David Stenn, "It Happened One Night . . . at MGM," *Vanity Fair*, April 2003, http://www.vanityfair.com/fame/features/2003/04/mgm200304.

122 Ibid.

123 Ibid.

124 Ibid.

125 For example: "Miss Douglas, who lives at 1160 Bronson Avenue, charges that she was induced to attend the party." "Girl Identifies Suspect in Film Barn Party Attack," *Los Angeles Times*, June 15, 1937.

126 Stenn, "It Happened One Night."

127 Davis, *Glamour Factory*, 93.

128 Stenn, "It Happened One Night."

129 Marion, *Off with Their Heads*, 164.

130 Davis, *Glamour Factory*, 306.

131 Ibid.

132 Peggy Montgomery, interviewed in *Girl 27*, directed by David Stenn, 2007.

133 Charles Schreger, "A Woman Leader of a Screen Union," *Los Angeles Times*, June 23, 1980.

134 Douglas Martin, quoted in Margaret Sullivan, "Gender Questions Arise in Obituary of Rocket Scientist and Her Beef Stroganoff," *New York Times*, April 1, 2013, http://publiceditor.blogs.nytimes.com/2013/04/01/gender-questions-arise-in-obituary-of-rocket-scientist-and-her-beef-stroganoff/?_r=0.

Chapter 4 "His Acolyte on the Altar of Cinema": The Studio Secretary's Creative Service

1 "Enter 'Social Mentor,'" *Moving Picture World* (hereafter *MPW*) 25, no. 8 (August 21, 1915): 1303.

2 Citing Ivan Illich, Arlie Hochschild describes shadow labor as "an unseen effort, which, like housework, does not quite count as labor but is nevertheless crucial to getting things other things done." Hochschild, *The Managed Heart: Commercialization of Human Feeling* (Berkeley: University of California Press, 2003), 167.

3 "Secretary to Executive Assistant—Job Description," University of Southern California Warner Bros. Archive, Personnel Files Box 382B, Undated.

4 Miriam Simons Leuck, *Fields of Work for Women* (New York: D. Appleton and Company, 1926), 51; emphasis added.

5 Hochschild, *The Managed Heart*, 9.

6 Ibid., 147.

7 Ibid., 163.

8 "Affirms," ibid., 165; "Creating," ibid., 20.

9 Ibid., 164.

10 Ibid., 20.

11 Rae Chatfield Ayer, "Are Men Better Secretaries? No!" *Rotarian* (November 1940): 58. Quoted in Lynn Peril, *Swimming in the Steno Pool* (New York: W. W. Norton, 2011), 32.

12 Ibid., quoting Ruth McKay, "White Collar Girl," *Chicago Tribune*, April 20, 1943, 18.

13 Mervyn LeRoy, *It Takes More Than Talent* (New York: Alfred A. Knopf, 1953), 145–147.

14 Ibid., 146.

15 Ibid., 153.

16 Though Ann Richards and Faith Whittlesey have been mistakenly credited with this quote, it was written by cartoonist Bob Thaves in his comic strip. Bob Thaves, *Frank and Ernest* cartoon, unidentified date, 1982.

17 Particularly domestic labor, which is often not recognized as having economic value at all.

18 Organisation for Economic Co-operation and Development, *The Future of Female-Dominated Occupations* (Paris: Author, 1998), 51.

19 Leo Rosten, *The Movie Colony, the Movie Makers* (New York: Arno Press, 1941), 32.

20 Valeria Beletti, letter to Irma Prina, December 1, 1924, in Cari Beauchamp, ed., *Adventures of a Hollywood Secretary: Her Private Letters from Inside the Studios of the 1920s* (Berkeley: University of California Press, 2006), 12.

21 Valeria Beletti, letter to Irma Prina, May 18, 1926, in Beauchamp, ed., *Adventures of a Hollywood Secretary*, 136.

22 Samuel Goldwyn, letter of recommendation for Valeria Beletti, July 22, 1926, in Beauchamp, ed., *Adventures of a Hollywood Secretary*, 147.

23 Valeria Beletti, letter to Irma Prina, February 19, 1925, in Beauchamp, ed., *Adventures of a Hollywood Secretary*, 16.

24 Valeria Beletti, letter to Irma Prina, May 20, 1925, in Beauchamp, ed., *Adventures of a Hollywood Secretary*, 45.

25 Valeria Beletti, letter to Irma Prina, May 6, 1925, in Beauchamp, ed., *Adventures of a Hollywood Secretary*, 43–44.

26 Valeria Beletti, letter to Irma Prina, June 26, 1925, in Beauchamp, ed., *Adventures of a Hollywood Secretary*, 57.

27 Valeria Beletti, letter to Irma Prina, April 8, 1925, in Beauchamp, ed., *Adventures of a Hollywood Secretary*, 32.

28 Valeria Beletti, letter to Irma Prina, July 14, 1925, in Beauchamp, ed., *Adventures of a Hollywood Secretary*, 72.

29 Valeria Beletti, letter to Irma Prina, May 1, 1926, in Beauchamp, ed., *Adventures of a Hollywood Secretary*, 133.

30 Valeria Beletti, letter to Irma Prina, September 5, 1925, in Beauchamp, ed., *Adventures of a Hollywood Secretary*, 73.

31 Valeria Beletti, letter to Irma Prina, July 20, 1926, in Beauchamp, ed., *Adventures of a Hollywood Secretary*, 146–147.

32 Valeria Beletti, letter to Irma Prina, June 4, 1926, in Beauchamp, ed., *Adventures of a Hollywood Secretary*, 138.

33 Valeria Beletti, letter to Irma Prina, July 15, 1926, in Beauchamp, ed., *Adventures of a Hollywood Secretary*, 144.

34 Beauchamp, ed., *Adventures of a Hollywood Secretary*, 145.

35 Described in "Clark Gable Walked off the Set for Racism," *Filmbeat*, updated June 5, 2013, http://www.filmibeat.com/hollywood/news/2008/gable-racism-victor-fleming-300908.html, an account reportedly related by Michael Sragow from his book on Victor Fleming, *Victor Fleming: An American Master* (New York: Pantheon Books, 2008).

36 These are the dates Rabwin gives for her employment with Selznick, which began shortly after he was hired to head RKO, and ended when she resigned from Selznick International Pictures in 1944, but accounts of her tenure's duration vary. Rabwin also says she worked for Selznick for fifteen years at one point. Other accounts—a *Variety* article and David Thompson's Selznick biography, *Showman*, report that she "retired" for brief stints in the mid- to late 1930s, but they differ as to the dates. What seems clear is that Rabwin worked with Selznick from 1932 to at least 1934, when she wed her husband, and was present for much of the lead-up to and the production of *Gone with the Wind* from early 1937 until at least 1939 and was in her role as the producer's right hand during his collaboration with

Alfred Hitchcock in 1939 and 1940. Rabwin, *Yes, Mr. Selznick: Recollections of Hollywood's Golden Era* (Pittsburgh, PA: Dorrance, 1999), 165–168; "Selznick Aide Quits," *Daily Variety,* September 16, 1939, David Thomson, *Showman: The Life of David O. Selznick* (New York: Alfred A. Knopf, 1992), 212, 238.

37 Rabwin, *Yes, Mr. Selznick*, 163.

38 Ibid., 165.

39 Ibid., 165.

40 Peril, *Swimming in the Steno Pool*, 23.

41 Today, the word *assistant* has an even broader range of meanings in the American media industry. The male version of the executive assistant job does not exist, except perhaps a junior executive position. The title "executive assistant" may reflect a role that is elevated from that of assistant, or it may simply be the title a producer gives the person who assists him at the office, or even to the person who works out of her home, who might more accurately be labeled a personal assistant. I have also encountered personal assistants who were categorized as "production assistants" and even "executive assistants," even though they only handled personal matters. In the 1930s and 1940s, the terms *secretary* and *private secretary* encompassed a number of functions that varied by employer. Both then and now, job title and duties are somewhat idiosyncratic and dependent on the office culture, the employer, and so on.

42 "Gals and Gab," *Daily Variety*, March 3, 1937, 3.

43 Lucie Arnaz, "Foreword," *Yes, Mr. Selznick*, xi.

44 Rabwin, *Yes, Mr. Selznick*, 114.

45 Bob Thomas, *Selznick* (Garden City, NY: Doubleday, 1970), 66.

46 Marcella Rabwin, *Gone with the Wind Newsletter* (November 1988), 1.

47 Thomas, *Selznick*, 66.

48 Rabwin, *Yes, Mr. Selznick*, 162.

49 Marcella Rabwin, interview in "Hitchcock, Selznick, and the End of Hollywood," *PBS American Masters*, Episode 1.14, Michael Epstein, dir., 1999.

50 Thomas, *Selznick*, 239.

51 David Thomson, interview in "Hitchcock, Selznick, and the End of Hollywood."

52 "Brain-damaged" in Thomson, *Showman*, 168; "patching things up" in Rabwin, *Yes, Mr. Selznick*, 167.

53 Rabwin, *Yes, Mr. Selznick*, 167. Also discussed in Thomas, *Selznick*, 125: "Waiting for an appointment with the boss could be a nerve-wracking ordeal. The employee was told by Marcella Rabwin, 'stand by, Mr. Selznick wants to see you'" and then made to wait.

54 Rabwin, *Gone with the Wind Newsletter*.

55 "The Women Who Run the Men: Secretaries to Hollywood's Film Chiefs Are Key Links in the Industry," *Variety*, October 7, 1942, 15.

56 Robert Parrish, *Growing Up in Hollywood* (New York: Harcourt Brace Jovanovich, 1976), 97.

57 Rabwin, *Yes, Mr. Selznick*, 166–167.

58 Thomas, *Selznick*, 66.

59 Rabwin interview in "Hitchcock, Selznick, and the End of Hollywood."

60 Rabwin, *Yes, Mr. Selznick*, xiii.

61 Ibid., 87.

62 Ibid., 81.

63 Rabwin interview in "Hitchcock, Selznick, and the End of Hollywood."

64 Rabwin, *Yes, Mr. Selznick*, 85.

65 Ibid., 149–150, 161.

66 Ibid., 149.

67 Lois Hamby, interview in "Hitchcock, Selznick, and the End of Hollywood." For Schulman's book see Silvia Schulman (Jane Allen pseud.), *I Lost My Girlish Laughter* (New York: H. Wolf, 1938).

68 Thomson, *Showman*, 199. Move also discussed in "Selznick's 10 for UA," *Daily Variety,* June 19, 1935, 8.

69 For example, in the 1970s, when she wrote a recommendation of the Selznick-produced *David Copperfield* (1935) in the *LA Times* "Calendar" section, she added, "I was executive assistant to Mr. Selznick for many, many years—so I should know." Marcella Rabwin, "Getting Dickens Straight," *LA Times Calendar,* September 12, 1976, 2. Conferences discussed in "Marcella Bannett Rabwin— Obituaries" *Daily Variety,* December 31, 1998, 9.

70 Rabwin, *Yes, Mr. Selznick*, 166.

71 In his Selznick biography, Bob Thomas summarizes Rabwin's resignation by saying, "The press of her own family duties prompted Mrs. Rabwin to leave the Selznick organization."

72 Robert Vogel, *Robert Vogel Oral History*, interviewed by Barbara Hall, Margaret Herrick Library, 1990, 101.

73 "Mrs. Ida Koverman, Film Leader, Dies," *Daily Variety,* November 25, 1954, B1.

74 "Stars in her eyes" in Vogel, *Robert Vogel Oral History*, 101; "anxious to leave" in Finch and Rosenkrantz, *Gone Hollywood*, 254.

75 "Close friendship" in Bosley Crowther, *Hollywood Rajah: The Life and Times of Louis B. Mayer* (New York: Holt, 1960), 127–128; Meetings with Hearst/Hoover discussed in ibid., 136–137, 146.

76 Mike Connolly, "Just for Variety," *Daily Variety* (July 31, 1951), 2; see also "Ida Koverman—Obituary," *Variety* (December 1, 1954), 79.

77 Davis, *The Glamour Factory*, 20.

78 "The Women Who Run the Men," *Variety*, 15.

79 Steven Bingen, Stephen X. Sylvester, and Michael Troyen, *M-G-M: Hollywood's Greatest Backlot* (Solana Beach, CA: Santa Monica Press, 2011), 27.

80 Ibid.

81 Vogel, *Robert Vogel Oral History*, 102.

82 "Mrs. Ida Koverman, Film Leader, Dies," *Daily Variety* (November 25, 1954), B1.

83 Beth Day, *This Was Hollywood* (New York: Doubleday, 1960), 70.

84 Christopher Finch and Linda Rosenkrantz, *Gone Hollywood: The Movie Colony in the Golden Age* (New York: Doubleday, 1979), 254.

85 The executive secretary was also a formidable bookkeeper. Her background in finance gave her the acumen to reorganize the studio's accounting system. Ibid.

86 Ibid.

87 Beauchamp, *Without Lying Down*, 248.

88 "Inspiration in Her Life Story," *MGM Studio Club News* (November 14, 1938): 3, 10.

89 Beauchamp, *Without Lying Down*, 248.

90 Marion, *Off With Their Heads: A Serio-Comic Tale of Hollywood*, (New York: Macmillan, 1972), 199–200.

91 "Hundreds at Rites for Ida Koverman," *Los Angeles Times* (November 27, 1954), 7.

92 For dress codes see also Finch and Rosenkrantz, *Gone Hollywood*, 254. "studio mother" from Koverman's description of Green Room in: Ida R. Koverman, "The Little Green Room," *MGM Studio News* (July 1937).

93 Day, *This Was Hollywood*, 74.

94 "If there was a question" quote from Beauchamp, *Without Lying Down*, 248; "both at his home" quote from Norman Zierold, *The Moguls: Hollywood's Merchants of Myth* (Los Angeles: Silman-James, 1991), 309–310.

95 Ibid; see also Day, *This Was Hollywood*, 71.

96 Day, *This Was Hollywood*, 310.

97 Finch and Rosenkrantz, *Gone Hollywood*, 254.

98 "One of the most influential" in Day, *This Was Hollywood*, 71; see also Zierold, *The Moguls*, 320.

99 "They've laughed long enough" in Day, *This Was Hollywood*, 84; Hedda Hopper, "Hedda Hopper's Hollywood," *Los Angeles Times* (March 19, 1938), A17.

100 Day, *This Was Hollywood*, 84.

101 David Shipman, *Judy Garland: The Secret Life of an American Legend* (New York: Hyperion, 1994), 45.

102 Though Robert Vogel, Marcella Rabwin, and Frances Marion all recount some version of the story of Koverman auditioning Judy, it also appears in James Robert Parish and Michael R. Pitts, *Hollywood Songsters: Garland to O'Connor* (London: Taylor & Francis, 2003), 335.

103 Quoted in Randy L. Schmidt, *Judy Garland on Judy Garland: Interviews and Encounters* (Chicago: Chicago Review Press, 2014), 238.

104 Shipman, *Judy Garland*, 45.

105 Marion, *Off with Their Heads*, 198. Hugh Fordin states that whenever Koverman would bring in one of her protégés, she advised them to sing the Jewish traditional "Kol Nidre," and that "It was a known fact that this inevitably would lead, at least, to a short-term contract." In *M-G-M's Greatest Musicals: The Arthur Fried Unit* (Boston: Da Capo Press, 1995), 5.

106 Ibid.

107 Zierold, *The Moguls*, 300.

108 Day, *This Was Hollywood*, 84.

109 Sex appeal discussed in Zierold, *The Moguls*, 300. "Landslide vote" in Day, *This Was Hollywood*, 84.

110 "Mrs. Ida Koverman, Film Leader, Dies," *Daily Variety,* November 25, 1954, B1.

111 "The Women Who Run the Men," *Variety*, 15.

112 Marion, *Off with Their Heads*, 228.

113 Vogel, *Robert Vogel Oral History*, 102. Murphy quoted in "Hundreds at Rites for Ida Koverman," *Los Angeles Times,* November 27, 1954, 7.

114 Ida Koverman, "Nothing to Do 'Til Tomorrow," *MGM Studio Club News* (May 1941): 11.

115 "Meta Wilde—Obituary," *Times* [London], November 3, 1994.

116 Meta Carpenter Wilde and Orin Borsten, *A Loving Gentleman: The Love Story of William Faulkner and Meta Carpenter* (New York: Simon and Schuster, 1976), 18.

117 Ibid., 34.

118 Ibid., 177.

119 Ibid., 43.

120 Ibid., 36.

121 Ibid., 37–38.

122 Ibid., 38.

123 Ibid., 106, 112.

124 Ibid., 92.

125 Ibid.

126 Ibid., 155.

127 Ibid.

128 Ibid., 175.

129 Ibid., 176.

130 Ibid., 247–248.

131 Ibid., 36.

132 Ibid., 49.

133 Ibid., 25.

134 Ibid., 29.

135 Ibid., 84.

136 Ibid., 84–85.

137 Ibid., 264.

138 Ibid., 288.

139 Joel Williamson, *William Faulkner and Southern History* (Oxford: Oxford University Press, 1995), 248–249.

140 Wilde and Borsten, *A Loving Gentleman*, 9–10.

141 Ibid., 10.

142 Ibid., 108.

143 Ibid., 331.

144 Peggy Robertson, *Peggy Robertson Oral History*, interview by Barbara Hall, Margaret Herrick Library, 1995.

145 Reville, Robertson, and their significant contributions to Hitchcock's films take center stage in the Sacha Gervasi-directed *Hitchcock* (Fox Searchlight, 2012).

146 Myrna Oliver, "Peggy Robertson: Personal Assistant to Alfred Hitchcock—Obituary," *Los Angeles Times,* February 12, 1998, http://articles.latimes.com/1998/feb/12/news/mn-18394.

147 Robertson, *Peggy Robertson Oral History*, 86.

148 John Houseman told Taylor, "I would put my hand in the fire to swear [Harrison] was never his mistress. I ought to know, because for some time she was mine." All quotes from John Russell Taylor, "The Truth about Hitch and Those Cool Blondes," *Times,* April 5, 2005, 2, 4.

149 Robertson, *Peggy Robertson Oral History*, 303.

150 Ronald Bergan, "Obituary: Peggy Robertson: Smoothing Out the Hitch," *Guardian*, February 16, 1998, 13.

151 Ibid. See also Robertson, *Peggy Robertson Oral History*, 271.

152 Robertson, *Peggy Robertson Oral History*, 129.

153 For examples see *The Birds*, Folder 6f. 64 "Corr. re: Tipi Hedren Trip to Europe," Margaret Herrick Library Special Collections, 1963.

154 Robertson, *Peggy Robertson Oral History*, 265.

155 Ibid., 266.

156 Ibid., 256.

157 Ibid., 100–101.

158 Ibid., 199.

159 Ibid., 225.

160 Ibid.

161 Ibid., 225–226.

162 Ibid., 231.

163 Ibid., 276.

164 "Bogdanovich Planning Six Features, Four in Texas," *Daily Variety*, July 9, 1982, 21; Oliver, "Peggy Robertson."

165 Kristyn Burtt, "The Obsession with His Film Muse," *Sheknows.com*, October 19, 2012, http://www.sheknows.com/entertainment/articles/974433/tippi-hedren-reveals-the-real-hitchcock.

166 Tim Oglethorpe, "Hitchcock? Was He a Psycho?" *London Daily Mail*, December 21, 2012. 58.

167 Ronald Bergan, "Obituary: Peggy Robertson: Smoothing Out the Hitch," 13.

168 Marx, *A Gaudy Spree*, 10.

169 Idwal Jones, "The Muse in Hollywood," *New York Times*, December 27, 1936, X4.

170 Valeria Beletti, letter to Irma Prina, February 1926, in Beauchamp, ed., *Adventures of a Hollywood Secretary*, 116.

171 Letter to Irma Prina, November 1927, in Beauchamp, ed., *Adventures of a Hollywood Secretary*, 166–167.

Chapter 5   Studio Girls:
Women's Professions in Media Production

1 This approach is modeled after John Caldwell's use of trade stories as deep texts to be mined for evidence of industrial practices. *Production Culture: Industrial Reflexivity and Critical Practice in Film and Television* (Durham, NC: Duke University Press, 2003).

2 Marcia McCreadie, *The Women Who Write the Movies: From Frances Marion to Nora Ephron* (New York: Birch Lane Press, 1994), 149.

3 Ibid., 152.

4 Such claims were often supported by invocations of women's success as fiction writers. However, as Susan Coultrap-McQuinn explains, women often succeeded through the guidance of male publishers, who began to see them as

"unprofessional" as their businesses grew and reorganized along efficient lines.
For one example of the comparison, see interview with Clara Berenger, "Feminine
Sphere in the Field of Movies Is Large Indeed," *Moving Picture World* (hereafter
referred to as *MPW*) (August 2, 1919): 662. Susan Coultrap-McQuinn, *Doing
Literary Business: American Women Writers in the Nineteenth Century* (Chapel
Hill: University of North Carolina Press), 1990, 48.

5 Alice Guy-Blaché, "Woman's Place in Photoplay Production," *MPW* 21, no. 2
(July 11, 1914): 195.

6 "Sleeping soul" in: June Mathis, "Harmony in Picture-Making," *Film Daily*, May 6,
1923, 5; "human quality" in June Mathis, "The Feminine Mind in Picture Making,"
Film Daily, June 7, 1925, 115.

7 Mathis, "The Feminine Mind," 115.

8 Florence M. Osbourne (ed.), "Why Are There No Women Directors?" *Motion
Picture Magazine* (November 1925): 5.

9 Jane Murfin, "Sex and the Screen," in *The Truth about Movies by the Stars*, ed.
Laurence A. Hughes (Hollywood: Hollywood Publishers, 1924), 459–460.

10 Cari Beauchamp, *Without Lying Down: Frances Marion and the Powerful Women
of Early Hollywood* (Berkeley: University of California Press, 1997), 41.

11 For researchers in guild see "Job Titles and Classifications," University of Southern
California Warner Bros. Archive, Personnel Files Box 380b, dated November 1949.
Typing requirements in Mervyn LeRoy, *It Takes More Than Talent* (New York:
Alfred A. Knopf, 1953), 204, 206.

12 Library work discussed in Mary A. Laselle and Katherine E. Wiley, *Vocations
for Girls* (New York: Houghton Mifflin, 1913), 66; *Fields of Work for Women*
(New York: D. Appleton and Company, 1926), 110; Catherine Filene, ed., *Careers
for Women: New Ideas, New Methods and New Opportunities—To Fit a New World*
(Boston: Houghton Mifflin, 1934), 389.

13 Dorothy Hawley Cartwright, "Their Business Is Looking Up," *Talking Screen* 1,
no. 5 (June 14, 1930): 64–66, 88.

14 See Antonia Lant's *The Red Velvet Seat: Women's Writing on the First Fifty Years
of Cinema* (London: Verso, 2006), 548–549, Reina Wiles Dunn in "Off-Stage
Heroines of the Movies," *Independent Woman* 13 (1934): 202; Filene, *Careers for
Women*, 432.

15 LeRoy, *It Takes More Than Talent*, 203–204.

16 Melvin M. Riddle, "From Pen to Silversheet II—Architecture, Decoration,
Research," *Photodramatist* 3, no. 8 (January, 1922): 37.

17 Samuel Marx, *A Gaudy Spree: Literary Hollywood When the West Was Fun*
(New York: Franklin Watts, 1987), 9–10. Tenure through 1950s in Steven Bingen,
Stephen X. Sylvester, and Michael Troyen, *M-G-M: Hollywood's Greatest Backlot*
(Solana Beach, CA: Santa Monica Press, 2011), 59.

18 LeRoy, *It Takes More Than Talent*, 205–205.

19 Carl Milliken, "Information Please," *Warner Club News* (June 1940): 3.

20 "Storing up data . . ." and "through voluminous masses" quotes from Riddle,
"From Pen to Silversheet II," 37; "hand-in-hand . . ." quote from Marx, *A Gaudy
Spree*, 9–10; "writers, wardrobe and make-up . . ." quote from Beth Day, *This*

Was Hollywood: An Affectionate History of Filmland's Golden Years (New York: Doubleday, 1960), 128–129.

21 LeRoy, *It Takes More Than Talent*, 205.

22 Ibid.

23 Howard Koch, *As Time Goes By: Memoirs of a Writer* (New York: Harcourt, 1979), 43.

24 Day, *This Was Hollywood*, 128–129.

25 Kate Corbaley, "Duties and Qualifications of the Scenario Reader," *Opportunities in the Motion Picture Business* 2 (Los Angeles: Photoplay Research Society, 1922), 61–62.

26 Ibid., 62.

27 Davis, *The Glamour Factory*, 162.

28 Marx, *A Gaudy Spree*, 151–153.

29 Davis, *The Glamour Factory*, 163; Rudy Behlmer, ed., *A Memo from David O. Selznick* (New York: Viking, 1972), 138.

30 Leo Rosten, *The Movie Colony, the Movie Makers* (New York: Arno Press, 1970), 32.

31 Marx, *A Gaudy Spree*, 10.

32 Ibid., 10–11.

33 Ibid., 10.

34 A Scenario Editor, "The Movies: A Colossus That Totters," *Bookman* 48, no. 6 (February 1919): 655.

35 Filene, *Careers for Women*, 432–33.

36 Women often wrote columns for the reading department in studio newsletters, e.g., Lillian Bergquist, "Readings from the Reading Department," *RKO Studio Club News* (November 1940), 9. The columns often mention more female readers by name than they do male readers, as in May of 1941 when MGM readers Jeannie Melton and Marge Thorson reported on covering materials at that studio (*MGM Studio Club News* [May 1941]) Other female MGM readers included Peg LeVino, Rosina Knowles, Mildred Haig, and Elizabeth Dickson (mentioned for wearing slacks), along with Dorothy Pratt, who "gave her readers a synopsis on the joys of ranching" in June ("The Reading Department," *MGM Studio Club News* [June 1941], 9). By October 1942, the Reading column reports increase in population: "With all the men folks tooling off to war, the file department is beginning to be feminized. Elaine Speed, Crolyn Asher and Helly Koretz are the latest glamour acquisitions to the department. It is reported that Earl Booth Enjoys his job of explaining the file system to them and that Dan MacNeill has a constant glint in his eye" (Ab Jackson Jr., "Reading," *MGM Studio Club News* [October 1942], 9); Marge Thorson veteran reader leaves reading dept. to do "heavy work" with the MPD OWI returned to take over Ed Hogan's desk ("The Reading Department," *MGM Studio Club News* [June 1948], 8).

37 LeRoy, *It Takes More Than Talent*, 210.

38 Ibid., 149, 155.

39 Valeria Beletti, letter to Irma Prina, November 1927, in Cari Beauchamp, ed., *Adventures of a Hollywood Secretary: Her Private Letters from Inside the Studios of the 1920s* (Berkeley: University of California Press, 2006), 166–167.

40 Marx, *A Gaudy Spree*, 27.

41 Frances Marion, *Off with Their Heads: A Serio-Comic Tale of Hollywood* (New York: Macmillan, 1972), 204–205.

42 Marx, *A Gaudy Spree*, 26.

43 Marion, *Off with Their Heads*, 204–205.

44 Marx, *A Gaudy Spree*, 27.

45 For example, 20th Century Fox Filmed Entertainment's story department employs twenty-five in-house readers for union wages, who cover material for all Fox divisions. Ted Dodd, Fox senior vice president of creative affairs, interview by author, Los Angeles, CA, March 4, 2014.

46 Hadley Davis, *Development Girl* (New York: Random House, 1999), 11.

47 Compensation and promotion prospects discussed further in this book's epilogue.

48 Based on my experience as a story analyst for fifteen years at what is considered decent pay for the field, I can attest to the fact that this mythology is false.

49 Clare Naylor and Mimi Hare, *The Second Assistant: A Tale from the Bottom of the Hollywood Ladder* (New York: Plume, 2005); Chris Dyer, *The Loves of a D-Girl: A Novel of Sex, Lies, and Script Development* (New York: Plume, 2005).

50 Lizzie Francke, *Script Girls: Women Screenwriters in Hollywood* (London: BFI, 1994), 6.

51 John H. Rathbun, "Motion Picture Making and Exhibiting," *Motography* 9, no. 8 (April 19, 1913): 278.

52 Janet Staiger, in *The Classical Hollywood Cinema: Film Style and Mode of Production to 1960*, by David Bordwell, Janet Staiger, and Kristin Thompson (New York: Columbia University Press, 1985), 152.

53 Ibid., 206.

54 Ibid.

55 Helen Starr, "Putting It Together," *Photoplay* 14, no. 2 (July 1918): 54.

56 June Mathis, "The Feminine Mind in Picture Making," *Film Daily,* June 7, 1925, 115.

57 Robertson, "Peggy Robertson Oral History," 55.

58 Morris Abrams, "Interview with Morris Abrams," *The Role of the Script Supervisor in Film and Television* (New York: Hastings House, 1986), 20–21.

59 Robertson, "Peggy Robertson Oral History," 55.

60 Ibid.

61 Alma Young, "Interview with Alma Young." Interview by Anthony Slide and Robert Gitt, Margaret Herrick Library, 1977, 2.

62 Ibid., 3.

63 Valeria Beletti, letter to Irma Prina, February 2, 1926, in Beauchamp, ed., *Adventures of a Hollywood Secretary*, 112. Meta Carpenter Wilde and Orin Borsten, *A Loving Gentleman: The Love Affair of William Faulkner and Meta Carpenter* (New York: Simon and Schuster, 1976), 38. Catalina Lawrence, interview by Mike Steen in Mike Steen, *Hollywood Speaks! An Oral History* (New York: G. P. Putnam and Sons, 1974), 345–346.

64 "The Question Box," *Warner Club News* (November 1939), 8.

65 LeRoy, *It Takes More Than Talent*, 298.

66 Robertson, "Peggy Robertson Oral History," 54–55.

67 Ibid., 148.

68 "Script Supervisor Has No Margin for Error," *Daily Variety,* October 25, 1977, 112.

69 Young, "Interview with Alma Young," 3.

70 Filene, *Careers for Women,* 433.

71 Robertson, "Peggy Robertson Oral History," 148.

72 Ibid.

73 Ibid., 148–149.

74 Fred and Jan Yager, *Career Opportunities in the Film Industry* (New York: Ferguson, 2009), kindle location1726 of 4360.

75 For example see "The Accused," script supervisor's notes, Paramount Pictures Production Records, File #3.F-29, dated April 17, 1948, Margaret Herrick Library, Special Collections, Los Angeles, CA.

76 "The Question Box," 8.

77 Lawrence, *Hollywood Speaks,* 350–352.

78 Kerry Lyn McKissick, interviewed in Helena Lumme and Mika Manninen, *Great Women of Film* (New York: Billboard, 2002), 79.

79 "Script Supervisor Has No Margin for Error," 112.

80 Robertson. "Peggy Robertson Oral History," Discussion of "anxiety," on 63, "hated" on 87.

81 Abrams, *The Role of the Script Supervisor,* 21.

82 Ibid.

83 Robertson. "Peggy Robertson Oral History," 53.

84 "Script Supervisor Has No Margin for Error," 112.

85 Ibid.

86 May Wale Brown, *Reel Life on Hollywood Movie Sets* (Riverside, CA: Ariadne, 1995), 111.

87 Ibid., 12, 20.

88 Young, "Interview with Alma Young," 13.

89 "I win a bet from Billie—the script girl," Agfa Ad, *Daily Variety,* October 2, 1941, 5.

90 "Script Supervisor Has No Margin for Error," *Daily Variety,* 112.

91 Robertson. "Peggy Robertson Oral History," 49.

92 Brown, *Reel Life,* 168.

93 Young, "Interview with Alma Young," 11.

94 Robertson. "Peggy Robertson Oral History," 52.

95 Ibid., 99.

96 For example, May Wale Brown attempted to support an emotional Judy Holliday, and Meta Carpenter chatted companionably to actor Fredric March's wife whenever she visited set. Brown, *Reel Life,* 56–57; Wilde and Borsten, *A Loving Gentleman,* 282, 276.

97 *The Question Box,* 8.

98 Wilde and Borsten, *A Loving Gentleman,* 121.

99 Ibid., 262–263.

100 Ibid., 313.

101 Wilde and Borsten, *A Loving Gentleman,* 271.

102 Brown, *Reel Life,* 18.

103 Young, "Interview with Alma Young," 3.

104 Wilde and Borsten, *A Loving Gentleman*, 304.

105 Ibid., 304. 1958 Charter date in "History," IATSE LOCAL 871, https://www.ialoca1871
.org/AboutUs/History.aspx; script clerks were excluded from membership in the
Screen Office Employees Guild when it was elected to represent clerical workers
who did their jobs from offices. See "Columbia Pictures Corporation, et al.,"
Cases Nos. R-2035 to R-044 inclusive, National Labor Relations Board, Decided
October 8, 1940.

106 Jack Hellman, "Light and Airy," *Daily Variety*, October 30, 1947, 6.

107 Steen, *Hollywood Speaks*, 350.

108 Katie Salen and Eric Zimmerman, *Rules of Play: Game Design Fundamentals*
(Cambridge, MA: MIT Press, 2004, 94).

109 Abrams, *The Role of the Script Supervisor*, 22.

110 Lawrence, *Hollywood Speaks*, 323.

111 Karen Ward Mahar, *Women Filmmakers in Early Hollywood* (Baltimore: Johns
Hopkins University Press, 2005), 200–201.

112 "The Film Editor, His Training and Qualifications," *Opportunities in the Motion
Picture Business* no. 3 (Los Angeles: Photoplay Research Society, 1922), 75–76.

113 Helena Lume, *Great Women of Film* (New York, Billboard, 2002).

114 Edward Dmytryk, *It's a Hell of a Life but Not a Bad Living* (New York: Times
Books, 1978), 35. Also discussed as a practice at Paramount in Davis, *The Glamour
Factory*, 289.

115 Dmytryk, *It's a Hell of a Life*, 36.

116 Lawrence, *Hollywood Speaks*, 353.

117 Davis, *The Glamour Factory*, 292.

118 Mahar, *Women Filmmakers*, 201.

119 Margaret Booth, interview by Rudy Behlmer, in "An Oral History with Margaret
Booth," Margaret Herrick Library, Los Angeles, CA, quoted in Mahar, *Women
Filmmakers*, 201.

120 Dmytryk, *It's a Hell of a Life*, 5.

121 Described as "Robust physique and nerve force" in Osbourne, "Why Are There
No Women Directors?," 5. "When a woman takes a megaphone in hand, the
eyes of the entire industry are focused upon her," discussed in June Mathis, "The
Feminine Mind in Picture Making," *Film Daily*, June 7, 1925, 115. "Unless you are
hardy and determined, the director's role is not for you," in Ida May Park, "Motion
Picture Work: The Motion Picture Director," in Filene, *Careers for Women*, 337.

122 Ralph Rosenblum and Robert Karen, *When the Shooting Stops . . . the Cutting
Begins* (1979; rpt. New York: Da Capo, 1986), 68.

123 Ibid.

124 Ibid., 68–69.

125 Davis, *The Glamour Factory*, 284.

126 Rosenblum and Karen, *When the Shooting Stops*, 69.

127 Barbara McLean, interviewed by Thomas R. Stempel, American Film Institute,
Los Angeles, December 22, 1971. Quoted in Davis, *The Glamour Factory*, 292.

128 Quoted in ibid., 284.

129 MacPherson, "Functions of the Continuity Writer," 30.

130 Fazan quoted in Mahar, *Women Filmmakers*, 201.

131 Mahar, *Women Filmmakers*, 201.

132 Ibid.

133 Dmytryk, *It's a Hell of a Life*, 18.

134 Ibid.; Mahar, *Women Filmmakers*, 201–202.

135 Davis, *The Glamour Factory*, 297.

136 Ibid., 288.

137 Mahar, *Women Filmmakers*, 201.

138 Dede Allen, interviewed by Patrick McGilligan in *Women and the Cinema*, ed. Karen Kay and Gerald Peary (New York: E. P. Dutton, 1977), 201–203.

139 For a detailed discussion of the location and configuration of what I label "planning" departments, see chapter 3.

140 Mertyl Gebhart, "Business Women in Film Studios," *Business Women* (December 1923).

141 Adam Hull Shirk, "Breaking In to the Publicity End of Pictures," *Opportunities in the Motion Picture Business* no. 2 (Los Angeles: Photoplay Research Society, 1922), 57.

142 Ibid.

143 Davis, *The Glamour Factory*, 138–139.

144 Ibid.

145 Vogel, "Robert Vogel Oral History," interviewed by Barbara Hall, Margaret Herrick Library, 1990, 101.

146 Christopher Finch and Linda Rosenkrantz, *Gone Hollywood* (New York: Doubleday, 1979), 279.

147 This messiness is the subject of John Caldwell's forthcoming book *Para-Industries*. He provides an extended abstract and a summary of some of the work's essential tenets in "Para-Industry: Researching Hollywood's Blackwaters," *Cinema Journal* 52, no. 3 (Spring 2013): 157–165.

148 Neither the 399 Teamsters nor the Casting Society of America (CSA), of which most casting directors are members, collect data on the gender breakdowns of their membership, but as of March 2012, the CSA directory lists 131 male members (15 casting associates and 116 casting directors) and 390 female members (42 casting associates and 348 casting directors), which translates to 74.8 percent (ambiguous names were verified against IMDB.com listings and casting company websites). This figure includes a small number of overseas casting directors and excludes the reportedly small number of casting directors unaffiliated with the CSA, nor does it include commercial casting directors, reality casting producers, and casting agents, which function as completely separate, distinct professions from that of casting director. *Casting Society of America*, "Find a Member Page," http://www.castingsociety.com/.

149 Marcia Ross, CSA, interview by author, Los Angeles, CA, December 8, 2004.

150 "Always been a woman's field" quote from Meg Liberman, CSA, in Anne Berman, "Shattering the Casting Couch Myth: Contrary to Popular Notions, Women Dominate the Field," *Variety: Women in Showbiz Issue*, November 18, 2002, A8; "A field that women did well in" quote from Wallis Nicita, CSA, in Vernon Scott,

"Everyone, at Times, Loves a Casting Director," *UPI Hollywood Reporter*, August 15, 1987.

151 "It used to be the secretaries . . ." quote from Deb Manwiller, CSA, interviewed in Catherine Seipp, "Casting Directors Can Make You a Star," *UPI Press International* via COMTEX, June 18, 2003, 1008169w7095; "We can type up a deal . . ." quoted in Jane Jenkins, CSA, interviewed in "Casting Directors: Under their Expert Eyes, Aspirants Go from Glossies to Glory," *People Weekly*, March 27, 1991, 71.

152 As outlined by Judith Butler, under which gender is constructed by the individual in a tacit agreement with society to "sustain discrete and polar genders" by means of "a stylized repetition of acts, product of continuous reiterated acts of performance." Judith Butler, *Gender Trouble: Feminism and the Subversion of Identity* (London: Routledge, 1999), 179.

153 Juliet Taylor, CSA, quoted in Don Shewey, "They Comb New York to Give Its Movies a Special Look," *New York Times*, April 11, 1982, 19.

154 Sharon Bialy, CSA, interview by author, Los Angeles, CA, December 3, 2004 (italics mine).

155 Ibid.

156 Taylor, quoted in Shewey, "They Comb New York," 19.

157 "That actors" in Taylor, quoted in Shewey, "They Comb New York," 19; "That everybody" in Justine Baddeley, CSA, quoted in "Roundtable: Casting Directors," *Hollywood Reporter*, hollywoodreporter.com, December 7, 2006.

158 "Keeping. . ." in Kim Davis-Wagner, CSA, quoted in "Roundtable."; "You might have . . ." in Ross, interview.

159 "It's almost . . ." in Bialy, interview; "Sometimes . . ." in Taylor, quoted in Shewey, "They Comb New York," 19; "Introducing . . ." Bialy, interview; "Keeping the room . . ." in Debi Manwiller, CSA, interview by author, Los Angeles, CA, December 8, 2004.

160 "The better casting" in Sherry Thomas, CSA, interview by author, Los Angeles, CA, December 3, 2004; "casting directors who" in Ross, interview.

161 "We're very emotional . . ." in Cathy Sandrich, C.S.A., interview by author, Los Angeles, December 7, 2004; "You have to tell . . ." in Ross, interview; "It can get you down . . ." in Nicita, quoted in Scott, "Everyone, at times . . ."; "There's a politic . . ." in Manwiller, interview by author.

162 Reality casters look not for performance skills in actors, but for the real "personalities" or "types" around which shows' reality narratives will be built. Workers are thus part casting director, but also producer, salesperson, and psychologist. Vicki Mayer, *Below the Line: Producers and Production Studies in the New Television Economy* (Durham, NC: Duke University Press, 2011), 134–135.

163 Mike Fenton, CSA quoted in Scott, "Casting Directors."

164 Ross, interview.

165 Bialy, interview.

166 Pitching services described in Bialy, interview; Scott, "Casting Directors"; Other skills in Liroff, CSA. "Inside . . . Part 1." Ross, interview.

167 Melvin M. Riddle, "From Pen to Silversheet VI—Casting the Characters," *Photodramatist* 3, no. 12 (May 1922): 25–26.

168 Staiger, *The Classical Hollywood Cinema*, 149.

169 Charles Graham, quoted in Pamela Robertson Wojcik, "Typecasting," *Criticism* 45, no. 2 (Spring 2003): 227–237.

170 "Premier Simplifies Casting," *Motography* 14, no. 25 (December 11, 1915): 1284.

171 "Equitable's Casting Director," *Motography* 14, no. 21 (November 20, 1915): 1075.

172 Mary Pickford, quoted in *Cinema: A Practical Course in Cinema Acting* (London: Standard Art Book Co., 1919), 29.

173 Staiger, *The Classical Hollywood Cinema*, 149.

174 Robert B. McIntyre, "How the Casting Director Selects Faces, Forms, and Types," *Opportunities in the Motion Picture Business* no. 1 (Los Angeles: Photoplay Research Society, 1922): 65.

175 Melvin M. Riddle, "From Pen to Silversheet VI—Casting the Characters," *Photodramatist* 3, no. 12 (May 1922): 25–26.

176 Ibid.

177 H. O. Davis, "A Kitchener among Cameras," *Photoplay* 11, no. 6 (May 1917): 129–131, 147, 168–169.

178 Quoted in: Riddle, "From Pen to Silversheet VI," 25–26.

179 Anthony Slide, *The New Historical Dictionary of the American Film Industry* (1998; reprint, London: Routledge, 2013), 35.

180 Tino Balio, *Grand Design Hollywood as a Modern Business Enterprise, 1930–39* (Berkeley: University of California Press, 1993), 145.

181 Ibid., 112.

182 Davis, *The Glamour Factory*, 91.

183 Ibid.

184 Ruth Burch, interviewed by Mike Steen in Steen, *Hollywood Speaks*, 358.

185 Ezra Goodman, "How to Be a Hollywood Producer," *Harper's,* May 1948, 418–419.

186 Day, *This Was Hollywood*, 112.

187 Henry Mintzberg, *The Nature of Managerial Work* (New York: Harper & Row, 1973), 169.

188 For further discussion of the gender makeup of clerical staffs in casting and other planning departments, as well as studio hierarchy and geography, see chapter 3. For examples of studio hierarchy see "RKO Studio Organization Chart from 1934," reproduced in Richard B. Jewell, *The Golden Age of Cinema: Hollywood 1929–1945* (Oxford: Blackwell, 2007), 67.

189 Davis, *The Glamour Factory*, 42.

190 Finch and Rosenkrantz, *Gone Hollywood*, 224–225.

191 James S. Ettema, "The Organizational Contexts of Creativity," in *Individuals in Mass Media Organizations: Creativity and Constraint* (London: Sage, 1983), 91.

192 Mike Fenton, CSA, quoted in Janet Maslin with Martin Kasindorf, "Finders Keepers," *Newsweek*, March 14, 1977, 92.

193 Billy Hopkins, CSA, interviewed in Glenn Collins, "For Casting, Countless Auditions and One Couch, Never Used," *New York Times*, January 30, 1990, C15.

194 Marci Liroff, "Inside the World of Casting, Part 1," *Facebook* note, April 4, 2011, http://www.facebook.com/note.php?note_id=10150158264996721.

195 Fenton, quoted in Maslin and Kasindorf, "Finders Keepers," 92.

196 Ross, interview.

197 For one early example of a woman receiving promotion from secretary to assistant in casting, see "Film Secretary Promoted," *Los Angeles Times*, October 4, 1936, 4.

198 Burch was first employed in 1932 "as a secretary at the Hal Roach Studios. In a short while I became the personal secretary and assistant to Mr. Roach. He later made me the studio casting director." Burch, *Hollywood Speaks*, 354; Dougherty promotion discussed in Dan Georgakas and Kevin Rabalais, "Fifty Years of Casting: An Interview with Marion Dougherty," *Cineaste* 25, no. 1 (Spring 2000): 26.

199 Davis, *The Glamour Factory*, 80, 85.

200 Finch and Rosenkrantz, *Gone Hollywood*, 45–52.

201 Ibid., 85–86.

202 Ibid., 132.

203 Day, *This Was Hollywood*, 115–116.

204 David A. Cook, *Lost Illusions: American Cinema in the Shadow of Watergate and Vietnam, 1970–1979* (Berkeley: University of California Press, 2002), 21–22.

205 Allen J. Scott, *On Hollywood: The Place, the Industry* (Oxford: Oxford University Press, 2005), 39.

206 Transition detailed in: Michael J. Bandler, "Casting Is His Lot," *American Way* (November 1982): 22. "CSA Searches for Respect, Identity in 'New' Hollywood," *Daily Variety*, June 26, 1989, I-32.

207 Shewey, "They Comb New York," 19.

208 Ibid., 26.

209 Burch, *Hollywood Speaks*, 355.

210 Described in: Joseph Turow, "Casting for TV Parts: The Anatomy of Social Typing," *Journal of Communication*, no. 4 (December 1978): 21.

211 Sam Christensen (casting director of *M*A*S*H* and other late 1970s and early 1980s television), interview with author, Studio City, CA, February 17, 2012.

212 Dougherty, "Fifty Years of Casting," 26.

213 Ibid.

214 Joyce Selznick, Interviewed by Joseph McBride, *Filmmakers on Filmmaking*, Vol. 2 (Los Angeles; Houghton Mifflin, 1983), 179.

215 Georgakas and Rabalais "Fifty Years . . . ," 26.

216 Nicita, quoted in Scott, "Everyone, at Times"; Taylor, in Shewey, "They Comb New York," 19.

217 Mollie Gregory, *Women Who Run the Show: How a Brilliant and Creative New Generation of Women Stormed Hollywood* (New York: St. Martin's Press, 2002), 11.

218 Berman, "Shattering," A8.

219 Bialy, interview.

220 Shift discussed in Mike Duffy, "Foreword," to Brian Wilson, *Soft Systems Methodology: Conceptual Model Building and Its Contribution* (New York: John Wiley & Sons, 2001), ix.

221 See Deborah Tannen, "The Sex-Linked Framing of Talk at Work," in *Gender and Discourse* (Oxford: Oxford University Press, 1994), 195–212.

222 Ross, interview.

223 Debra Zane, CSA, quoted in "Roundtable."

224 Ellen Lewis, CSA, ibid.

225 Bialy, interview.

226 Ross, interview.

227 Ibid.

228 Ibid.

229 Risa Bramon-Garcia, CSA quoted in Collins, "For Casting Directors." 15.

230 Jane Jenkins, quoted in "And the Oscar Doesn't Go To," *Backstage*, December 8, 2010, http://www.backstage.com/news/spotlight/and-the-oscar-doesnt-go-to/.

231 According to Mike Fenton, in the 1970s the going rate for casting services was $5,000–$7,000 per picture (an average of $400–$500 per week, which might be split between multiple casting directors and assistants). He observed that a member of the production crew responsible for fetching coffee and sweeping up made $8,400 ($600 per week) for the same period of time. In 1989, Fenton said, the casting director's rate was essentially unchanged. Meanwhile, the assistant director's pay had risen to $2,400 *per week*. "C.S.A. Searches," I-118–119.

232 Dougherty quoted by Taylor, in Shewey, "They Comb New York," 19. Steve Dyan, "Casting Directors Cast Their Fate with Teamsters," *Daily Variety*, June 24, 2005, 55.

233 Tracy Lillienfield, CSA, quoted in Berman, "Shattering," A8.

234 "C.S.A. Searches," I-118–119.

235 Unionization goals described in Will Tusher, "More American Actors Sought in O'Seas Pix," *Daily Variety*, September 21, 1983; for anti-union ad, see "Untitled Ad," *Daily Variety*, Tuesday, March 25, 1980, 32.

236 See: "C.S.A. Searches," I-119. Berman, "Shattering," A8.

237 Miranda Banks, "Gender below the Line," in *Production Studies: Cultural Studies of Media Industries*, ed. Miranda Banks, John Caldwell, and Vicki Mayer (London: Routledge, 2009), 95.

Epilogue: The Legacy of Women's Work in Contemporary Hollywood

1 Nikki Finke, "New Pay Schedule for WME Assistants," *Nikki Finke's Deadline Hollywood Daily,* July 29, 2009, http://www.deadline.com/2009/07/more-news-about-wme-assistant-pay/.

2 Ibid.

3 Ibid.

4 John Caldwell, *Production Culture: Industrial Reflexivity and Critical Practice in Film and Television* (Durham, NC: Duke University Press, 2008), 36.

5 Ibid., 34.

6 Ibid.

7 I base these claims primarily on interviews conducted with forty assistants in 2007, as well as my own experiences working as an assistant to TV and film producers for four years. Among assistants interviewed, salaries and work conditions ran the gamut from safe and respectful environments, with decent pay

and regular hours, to hostile, sexually inappropriate, or harassing workplaces, and workweeks of 80–100 hours without overtime. Detailed in Erin Hill, *Secretaries, Stenographers, and Assistants: A Report Funded by the California Women's Law Center* (Los Angeles, 2007), 71–111.

8 Bill Robinson and Ceridwen Morris, *It's All Your Fault: How to Make It as a Hollywood Assistant* (New York: Simon & Schuster, 2001), 14.

9 "Sorry Ari," *Entourage*, Julian Farino, dir., HBO (season 3, episode 12, 2006).

10 In salary comparisons between contemporary assistants and studio secretaries, the assistants' wages were equivalent to or lower than the secretaries when adjusted for inflation. Hill, *Secretaries, Stenographers, and Assistants*, 120.

11 Of the forty assistants I interviewed just a few years ago, only six have been promoted to junior executive positions, and in the same time, ten have left the industry completely. The rest have either moved to parallel sectors or are still working as assistants. Ibid.

12 Steven Greenhouse, "Judge Rules That Movie Studio Should Have Been Paying Interns," *New York Times,* June 11, 2013, http://www.nytimes.com/2013/06/12/business/judge-rules-for-interns-who-sued-fox-searchlight.html?_r=0.

Appendix

1 Stars and actors have been grouped together because it is difficult to distinguish between the two.

2 When possible, the exact number of workers visible is listed alongside specific types of labor. With larger groups or shots in which it is not possible to determine exact numbers, rough estimates have been given.

3 When specific workers are identified with specific jobs (for example, screenwriter) the specific job is listed. When it is not possible to determine which workers are doing which job in which department, the department is listed.

SELECTED BIBLIOGRAPHY

Abraham, Adam. *When Magoo Flew: The Rise and Fall of Animation Studio UPA*. Middletown, CT: Wesleyan University Press, 2012.

Abrams, Morris. *The Role of the Script Supervisor in Film and Television*. New York: Hastings House, 1986.

Anderson, Gregory, ed. *The White-Blouse Revolution: Women Office Workers since 1870*. Manchester: Manchester University Press, 1988.

Balio, Tino. *Grand Design: Hollywood as a Modern Business Enterprise, 1930–39*. Berkeley: University of California Press, 1993.

Banks, Miranda. "Bodies of Work: Rituals of Doubling and the Erasure of Film/TV Production Labor." PhD diss., University of California, Los Angeles, 2006.

Banks, Miranda, John Caldwell, and Vicki Mayer, eds. *Production Studies: Cultural Studies of Media Industries*. London: Routledge, 2009.

Bean, Jennifer M., and Diane Negra, eds. *A Feminist Reader in Early Cinema*. Durham, NC: Duke University Press, 2002.

Beauchamp, Cari, ed. *Adventures of a Hollywood Secretary: Her Private Letters from Inside the Studios of the 1920s*. Berkeley: University of California Press, 2006.

———. *Without Lying Down: Frances Marion and the Powerful Women of Early Hollywood*. Berkeley: University of California Press, 1997.

Becker, Howard. "An Epistemology of Qualitative Research." In *Ethnography and Human Development: Context and Meaning in Social Inquiry,* edited by R. Jessor, A. Colby, and R. A. Shweder, 53–71. Chicago: University of Chicago Press, 1996.

Bingen, Steven, Stephen X. Sylvester, and Michael Troyen. *M-G-M: Hollywood's Greatest Backlot*. Solana Beach, CA: Santa Monica Press, 2011.

Bodeen, DeWitt. *More from Hollywood! The Careers of 15 Great American Stars*. New York: A. S. Barnes and Company, 1977.

Bordwell, David, Janet Staiger, and Kristin Thompson. *The Classical Hollywood Cinema: Film Style and Mode of Production, to 1960*. New York: Columbia University Press, 1985.

Brown, May Wale. *Reel Life on Hollywood Movie Sets*. Los Angeles: Ariadne, 1995.

Brownlow, Kevin. *The Parade's Gone By. . . .* Berkeley and Los Angeles: University of California Press, 1977.

Butler, Judith. *Gender Trouble: Feminism and the Subversion of Identity*. 1990. Rpt. ed. London: Routledge, 1999.

Caldwell, John Thornton. "Para-Industry: Researching Hollywood's Blackwaters." *Cinema Journal* 52, no. 3 (Spring 2013): 157–165.

———. *Production Culture: Industrial Reflexivity and Critical Practice in Film and Television*. Durham, NC: Duke University Press, 2003.

Carman, Emily. "Independent Stardom: Freelance Stars and Independent Labor in 1930's Hollywood." PhD diss., University of California: Los Angeles, 2008.

Cerra, Julio Lugo, and Marc Wanamaker. *Movie Studios of Culver City*. Charleston, SC: Arcadia, 2011.

Checkland, Peter B. *Systems Thinking, Systems Practice*. Sussex: John Wiley & Sons, 1998.

Clark, Danae. *Negotiating Hollywood: The Cultural Politics of Actors' Labor*. Minneapolis: University of Minnesota Press, 1995.

Cook, David A. *Lost Illusions: American Cinema in the Shadow of Watergate and Vietnam, 1970–1979*. Berkeley: University of California Press, 2002.

Cooper, Mark Garrett. *Universal Women: Filmmaking and Institutional Change in Early Hollywood*. Chicago: University of Illinois Press, 2010.

Coultrap-McQuinn, Susan. *Doing Literary Business: American Women Writers in the Nineteenth Century*. Chapel Hill: University of North Carolina Press, 1990.

Crowther, Bosley. *Hollywood Rajah: The Life and Times of Louis B. Mayer*. New York: Holt, 1960.

Davies, Margery W. *Woman's Place Is at the Typewriter: Office Work and the Office Worker, 1870–1930*. Philadelphia: Temple University Press, 1982.

Davis, Ronald L. *The Glamour Factory: Inside Hollywood's Big Studio System*. Dallas, TX: Southern Methodist University Press, 1993.

Day, Beth. *This Was Hollywood: An Affectionate History of Filmland's Golden Years*. New York: Doubleday, 1960.

Dmytryk, Edward. *It's a Hell of a Life, but Not a Bad Living*. New York: Times Books, 1978.

Ettema, James. "The Organizational Contexts of Creativity." In *Individuals in Mass Media Organizations: Creativity and Constraint*. London: Sage, 1983.

Filene, Catherine, ed. *Careers for Women: New Ideas, New Methods and New Opportunities—To Fit a New World*. Boston: Houghton Mifflin, 1934.

Finch, Christopher, and Linda Rosenkrantz. *Gone Hollywood: The Movie Colony in the Golden Age*. New York: Doubleday, 1979.

Fine, Lisa. *The Souls of the Skyscrapers: Female Clerical Workers in Chicago, 1870–1930*. Philadelphia: Temple University Press, 1990.

Franke, Lizzie. *Script Girls: Women Screenwriters in Hollywood*. London: BFI, 1994.

Gaines, Jane. "Film History and the Two Presents of Feminist Film Theory." *Cinema Journal* 44, no. 1 (Fall 2004): 113–119.

———. "Pink-Slipped: What Happened to Women in the Silent Film Industry?" In *Blackwell's History of American Film*, edited by Roy-Michael Grundmann, Art Simon, and Cynthia A. Lucia, 000–000. London: Blackwell, 2012.

Gaines, Jane, and Radha Vatsal. "How Women Worked in the US Silent Film Industry." In *Women Film Pioneers Project*, edited by Jane Gaines, Radha Vatsal, and Monica Dall'Asta. Center for Digital Research and Scholarship. New York: Columbia University Libraries, 2013. https://wfpp.cdrs.columbia.edu/essay/how-women-worked-in-the-us-silent-film-industry/.

Gallagher, Katherine, and Stephen Greenblatt. *Practicing New Historicism*. Chicago: University of Chicago Press, 2000.

Gallon, Tom. *The Girl behind the Keys.* 1903. Rpt. ed. Arlene Young, ed. Toronto: Broadview, 2006.

Galtung, John. "A Structural Theory of Imperialism." *Journal of Peace Research* 8, no. 2 (1971).

Geertz, Clifford. *The Interpretation of Cultures: Selected Essays.* New York: Basic Books, 1977.

Georgakas, Dan, and Kevin Rabalais. In "Fifty Years of Casting: An Interview with Marion Dougherty." *Cineaste* 25, no. 2 (Spring 2000): 26.

Girl 27. Directed by David Stenn, 2007.

Gomery, Douglas. *The Hollywood Studio System: A History.* London: BFI, 2008.

Gone with the Wind: The Making of a Legend. Directed by David Hinton, 1988.

Gregory, Mollie. *Women Who Run the Show: How a Brilliant and Creative New Generation of Women Stormed Hollywood.* New York: St. Martin's Press, 2002.

Hallett, Hilary. *Go West, Young Woman! The Rise of Hollywood.* Berkeley: University of California Press, 2013.

Hastie, Amelie. *Cupboards of Curiosity: Women, Recollection, and Film History.* Durham, NC: Duke University Press, 2007.

Henderson, Felicia. "The Culture behind Closed Doors: Issues of Gender and Race in the Writers' Room." *Cinema Journal* 50, no. 2 (Winter 2011): 145–152.

Higham, Charles. *Cecil B. DeMille: A Biography of the Most Successful Film Maker of Them All.* New York: Charles Scribner's Sons, 1973.

Hill, Erin. "Re-Casting the Casting Director: Managing Change, Gendering Labor." In *Intermediaries: Management of Culture, Cultures of Management,* edited by Derek Johnson, Derek Kompare, and Avi Santo, 000–000. New York: New York University Press, 2014.

———. *Secretaries, Stenographers and Assistants: A Report Funded by the California Women's Law Center.* Los Angeles: California Women's Law Center, 2007.

———. "Women's Work: Femininity and Film and Television Casting." Console-ing Passions International Conference. Milwaukee, WI, 2006.

"Hitchcock, Selznick, and the End of Hollywood." Directed by Michael Epstein. *PBS American Masters,* Episode 1.14, 1999.

Hochschild, Arlie Russell. *The Managed Heart: Commercialization of Human Feeling.* 1983. Rpt. with new afterword. Berkeley: University of California Press, 2003.

Hochschild, Arlie, with Anne Machung. *The Second Shift.* New York: Penguin, 1989.

Holliday, Wendy. "Hollywood and Modern Women: Screenwriting, Work Culture, and Feminism, 1910–1940." PhD diss., New York University, 1995.

Jewell, Richard. *The Golden Age of Cinema: Hollywood, 1929–45.* London: Wiley Blackwell, 2007.

Kay, Karyn, and Gerald Peary, eds. *Women and the Cinema: A Critical Anthology.* New York: E. P. Dutton, 1977.

Kessler-Harris, Alice. *Out to Work: A History of Wage-Earning Women in the United States.* 1982. Rpt., 20th anniv. ed. Oxford: Oxford University Press, 2003.

Koszarski, Richard. *An Evening's Entertainment: The Age of the Silent Feature Picture, 1915–1928.* Berkeley: University of California Press, 1990.

Lant, Antonia, ed. *The Red Velvet Seat: Women's Writing on the First Fifty Years of Cinema.* London: Verso, 2006.

Laselle, Mary, and Katherine E. Wiley. *Vocations for Girls.* New York: Houghton Mifflin, 1913.

LeRoy, Mervyn. *It Takes More Than Talent.* New York: Alfred A. Knopf, 1953.

Leuck, Miriam Simons. *Fields of Work for Women.* New York: D. Appleton and Company, 1926.

Lewis, John, and Eric Smoodin, eds. *Looking Past the Screen: Case Studies in American Film History.* Durham, NC: Duke University Press, 2007.

Lume, Helena. *Great Women of Film.* New York: Billboard, 2002.

Lupton, Ellen. *Mechanical Brides: Women and Machines from Home to the Office.* New York: Princeton Architectural Press, 1993.

Mahar, Karen Ward. *Women Filmmakers in Early Hollywood.* Baltimore: Johns Hopkins University Press, 2006.

Marion, Frances. *Off with Their Heads: A Serio-Comic Tale of Hollywood.* New York: Macmillan, 1972.

Martindale, Hilda. *Women Servants of the State, 1830–1938.* Oxford: Oxford University Press, 1988.

Marx, Samuel. *A Gaudy Spree: Literary Hollywood When the West Was Fun.* New York: Franklin Watts, 1987.

Mayer, Vicki. *Below the Line: Producers and Production Studies in the New Television Economy.* Durham, NC: Duke University Press, 2011.

McBride, Joseph. *Filmmakers on Filmmaking.* Vol. 2. Los Angeles: Houghton Mifflin, 1983.

McCreadie, Marcia. *The Women Who Write the Movies: From Frances Marion to Nora Ephron.* New York: Birch Lane Press, 1994.

McGilligan, Patrick. *Backstory: Interviews with Screenwriters of Hollywood's Golden Age.* Berkeley: University of California Press, 1986, 36.

McHugh, Kathleen Anne. *American Domesticity: From How-to Manual to Hollywood Melodrama.* Oxford: Oxford University Press, 1999.

Musser, Charles. *Before the Nickelodeon: Edwin S. Porter and the Edison Manufacturing Company.* Berkeley: University of California Press, 1991.

Neal, Steven, ed. *The Classical Hollywood Reader.* London: Routledge, 2012.

Organisation for Economic Co-operation and Development. *The Future of Female-Dominated Occupations.* Paris: Author, 1998.

Ortner, Sherry. *Making Gender: The Politics and Erotics of Culture.* Boston: Beacon Press, 1996.

———. *Not Hollywood: Independent Film at the Twilight of the American Dream.* Durham, NC: Duke University Press, 2013.

Peril, Lynn. *Swimming in the Steno Pool.* New York: W. W. Norton, 2011.

Powdermaker, Hortense. *Hollywood: The Dream Factory.* New York: Little Brown, 1950.

Price, Leah, and Pamela Thurschwell, eds. *Literary Secretaries/Secretarial Culture.* Hampshire, UK: Ashgate, 2005.

Rabwin, Marcella. *Yes, Mr. Selznick: Recollections of a Golden Age.* Pittsburgh, PA: Dorrance 1999.

Robertson, Peggy. Peggy Robertson Oral History. Interview by Barbara Hall. Margaret Herrick Library, 1995.

Rosenblum, Ralph, and Robert Karen. *When the Shooting Stops . . . the Cutting Begins.* 1979. Rpt. ed. New York: Da Capo, 1986.

Rosten, Leo. *The Movie Colony, the Movie Makers.* New York: Arno Press, 1941.

Rothbauer, Paulette. "Triangulation." In *The SAGE Encyclopedia of Qualitative Research Methods,* edited by Lisa Given, 892–894. Thousand Oaks, CA: Sage.

Salen, Katie, and Eric Zimmerman. *Rules of Play: Game Design Fundamentals.* Cambridge, MA: MIT Press, 2004.

Schatz, Thomas. *The Genius of the System.* New York: Metro, 1996.

Scott, Allen. *On Hollywood: The Place, the Industry.* Princeton, NJ: Princeton University Press, 2005.

Seger, Linda. *When Women Call the Shots: The Developing Power and Influence of Women in Television and Film.* New York: Holt, 1996.

Senge, Peter M. *The Fifth Discipline: The Art and Practice of Learning Organization.* New York: Currency Doubleday, 1990.

Sigall, Martha. *Living Life inside the Lines: Tales from the Golden Age of Animation.* Jackson: University Press of Mississippi, 2005.

Simon, Joan, ed. *Alice Guy Blaché: Cinema Pioneer.* New Haven: Yale University Press, 2009.

Sito, Tom. *Drawing the Line: The Untold Story of Animation Unions from Bosko to Bart Simpson.* Lexington: University of Kentucky, 2006.

Sklar, Robert. *Movie-Made America: A Cultural History of American Movies.* New York: Vintage, 1994.

Slide, Anthony. *Early American Cinema.* Metuchen, NJ: Scarecrow, 1970.

———. *Early Women Directors.* New York: A. S. Barnes and Company, 1977.

———. *The Silent Feminists.* London: Scarecrow, 1996.

———. *Silent Topics.* London: Scarecrow Press, 2005.

Staiger, Janet. "Dividing Labor for Production Control: Thomas Ince and the Rise of the Studio System." *Cinema Journal* 18, no. 2 (1979): 16–25.

Stamp, Shelley. *Lois Weber in Early Hollywood.* Berkeley: University of California Press, 2015.

———. *Movie Struck Girls: Women and Motion Picture Culture after the Nickelodeon.* Princeton, NJ: Princeton University Press, 2000.

Steen, Mike. *Hollywood Speaks! An Oral History.* New York: G. P. Putnam and Sons, 1974.

Stenn, David. "It Happened One Night . . . at MGM." *Vanity Fair,* April 2003, http://www.vanityfair.com/fame/features/2003/04/mgm200304.

Stephens, E. J., and Marc Wanamaker. *Early Warner Bros. Studio.* Charleston, SC: Arcadia, 2010.

Strom, Sharon Hartman. *Beyond the Typewriter: Gender, Class, and the Origins of Modern American Office Work, 1900–1930.* Chicago: University of Illinois Press, 1992.

Tannen, Deborah. "The Sex-Linked Framing of Talk at Work." In *Gender and Discourse,* 195. Oxford: Oxford University Press, 1994.

Thomas, Bob. *Selznick.* Garden City, NY: Doubleday, 1970.

Thomson, David. Interview in "Hitchcock, Selznick, and the End of Hollywood." Directed by Michael Epstein. *PBS American Masters,* Episode 1.14, 1999.

———. *Showman: The Life of David O. Selznick*. New York: Alfred A. Knopf, 1992.

Turow, Joseph. "Casting for TV Parts: The Anatomy of Social Typing." *Journal of Communication* 4 (December 1978): 21.

Vasey, Ruth. *The World According to Hollywood*. Madison: University of Wisconsin Press, 1997.

Vidor, King. *King Vidor on Film Making*. New York: David McKay Company, 1972.

Vogel, Robert. Robert Vogel Oral History. Interview by Barbara Hall. Margaret Herrick Library, Los Angeles, CA, 1990.

Wasko, Janet. *Movies and Money: Financing the American Film Industry*. New York: Ablex, 1982.

Wilde, Meta Carpenter, with Orin Borsten. *A Loving Gentleman: The Love Story of William Faulkner and Meta Carpenter*. New York: Simon and Schuster, 1976.

Williamson, Joel. *William Faulkner and Southern History*. Oxford: Oxford University Press, 1995.

Wilson, Brian. *Soft Systems Methodology: Conceptual Model Building and Its Contribution*. New York: John Wiley & Sons, 2001.

Wojcik, Pamela Robertson. "Typecasting." *Criticism* 45, no. 2 (Spring 2003): 227–237.

Wood, Denis. *The Power of Maps*. New York: Guilford, 1992.

Yager, Fred and Jan. *Career Opportunities in the Film Industry*. New York: Ferguson, 2009.

Young, Arlene. "Introduction." In Grant Allen, *Typewriter Girl*, 1897. Broadview Encore Edition, 9–22. Toronto: Broadview Press, 2003.

Zierold, Norman. *The Moguls: Hollywood's Merchants of Myth*. Los Angeles: Silman-James, 1991.

Zohn, Patricia. "Coloring the Kingdom." *Vanity Fair*, March 2010, http://www.vanityfair.com/culture/features/2010/03/disney-animation-girls-201003.

INDEX

ABOUT THE AUTHOR

ERIN HILL worked in film and television production before pursuing study of the media industry. She lives in Los Angeles, where she continues freelance work in some of the very professions she researches. Hill is a visiting professor of cinema and media studies at the University of California, Los Angeles (UCLA) and in Dartmouth College's semester study program in Los Angeles, teaching courses on American film and television history, contemporary Hollywood, and production studies. She received the 2015 Society for Cinema and Media Studies dissertation award for the research on which this book is based. Her other work has appeared in *Making Media Work*, *The International Encyclopedia of Media Studies*, *Production Studies: Cultural Studies of Media Industries*, and *Reading Deadwood*. She earned her BA from the University of Michigan, and her MA and PhD from UCLA's School of Film, Television, and Digital Media.